Springer Series in Geomechanics and Geoengineering

Editors: Wei Wu · Ronaldo I. Borja

Ronaldo I. Borja (Ed.)

Multiscale and Multiphysics Processes in Geomechanics

Results of the Workshop on Multiscale and Multiphysics Processes in Geomechanics, Stanford, June 23–25, 2010

Springer

Professor Wei Wu, Institut für Geotechnik, Universität für Bodenkultur, Feistmantelstraße 4, 1180 Vienna, Austria, E-mail: wei.wu@boku.ac.at

Professor Ronaldo I. Borja, Department of Civil and Environmental Engineering, Stanford University, Stanford, CA 94305-4020, USA, E-mail: borja@stanford.edu

Editor

Ronaldo I. Borja
Department of Civil and Environmental Engineering
Stanford University
Stanford, California 94305-4020
E-mail: borja@stanford.edu

ISBN 978-3-642-19629-4 e-ISBN 978-3-642-19630-0

DOI 10.1007/978-3-642-19630-0

Springer Series in Geomechanics and Geoengineering ISSN 1866-8755

Library of Congress Control Number: 2011922616

Type Design and Cover Design: Scientific Publishing Services Pvt. Ltd., Chennai, India.

Printed on acid-free paper

5 4 3 2 1 0

springer.com

Foreword

Geomechanics is the study of the behavior of soils, rocks, and other geomaterials in their natural and engineered states. These materials exhibit a certain degree of permeability that allows fluids to enter into the inter-particle and/or inter-aggregate pores, to mix with each other, and to interact with the surrounding solids. The processes governing the interaction among the solid, liquid, and gas phases are highly complex and may involve mechanical, chemical, and electrical processes. Furthermore, the scale of interest may range from the nano-scale to the meter scale for material characterization purposes, and to hundreds of kilometers for geological and seismological applications.

This book is a compilation of the extended abstracts contributed by the participants at the International Workshop on Multiscale and Multiphysics Processes in Geomechanics held at Stanford University on June 23-25, 2010. It is by no means a complete compilation given the short duration of the workshop and the limited number of participants. Nevertheless, the articles in this book reflect some of the current trends in areas identified as being relevant to the theme of the workshop, namely, multiscale laboratory testing, constitutive modeling across different scales, multiphysics processes in porous materials, computational modeling, bridging of existing and new numerical tools, and the dynamics of fault zones. Accordingly, the abstracts are grouped into the following four sub-themes: experiments and constitutive models, multiscale processes, faults and discontinuities, and applications and simulations.

I would like to thank all the participants for their outstanding presentations and contributions; my workshop co-organizers Eric M. Dunham, Ellen Kuhl, and Joshua A. White; administrative associates Racquel Hagen and Kim Vonner; and the graduate students in the Structural Engineering and Geomechanics Program at Stanford University who helped ensure the success of the workshop. Finally, I would like to thank the U.S. National Science Foundation, Geomechanics and Geomaterials Program, for providing travel fellowships to young non-Stanford participants, and the John A. Blume Earthquake Engineering Center for sponsoring the local participants of the workshop.

19 February 2011
Stanford, California

Ronaldo I. Borja

Contents

Experiments and Constitutive Models

Multiscale Processes

EXPERIMENTAL EVIDENCE OF STRUCTURAL DEVELOPMENT INSIDE SHEAR BANDS IN SANDS

Amy L. Rechenmacher*, Sara Abedi, and Igor Faoro

Sonny Astani Department of Civil and Environmental Engineering
University of Southern California
3620 S. Vermont Ave, Los Angeles, CA 90089
* e-mail: arechenm@usc.edu

Summary. We show experimental evidence of vortices emanating from collapsed force chains in shear bands in real sands. From digital image-based local displacement data, a systematic, spatial pattern in kinematic data fields appears at the softening-critical state transition. By subtracting a linear shear displacement field from the observed non-affine displacement field, vortices appear and coordinate with the observed kinematics. We note differences in the vortex patterns observed in our sands from those observed in simulations on "ideal" granular materials, and discuss the fate of these vortices.

Keywords: force chain, length scale, shear band, vortex.

1 INTRODUCTION

Recent research has highlighted that the deformation field within shear bands in sands is non-uniform (e.g., [1–3]). Recent numerical efforts have further highlighted that these non-affine deformations are associated with vortex structures (e.g., [4,5]). Experimental capture of these vortices, in particular for real sands, is lacking and would serve to validate theoretical assumptions about the nature and length scales of these vortex structures.

Rechenmacher et al. [6] used digital imaging methods to track grain scale displacements within shear bands in sands deforming in plane strain. Various meso-scale kinematic measures along the length of the shear bands were quantified. At the global softening/critical state transition, a distinct spatial pattern among the kinematic quantities arose and was shown to be characteristic of a distinct, multi-force chain collapse event. Here, we show that this force chain collapse event is associated with the formation of vortices.

2 EXPERIMENTAL IMAGING AND DATA ANALYSIS

Strain-controlled, plane strain compression tests were conducted on dense sand specimens. Two different uniform, sieved mixtures of sand were used: silica-coarse (SC) sand had a median grain size of 0.40 mm, and masonry-coarse (MC) sand had a median grain size of 0.85 mm. The mixtures were designed for varied grain coloring for image-based displacement mapping. The base of the specimen rested on a ball-bearing "sled", and this lack of boundary constraint helped promote uninhibited shear band growth. High-resolution, experimental-grade digital images were captured throughout shear though glass walls that enforced the zero strain conditions.

R.I. Borja (Ed.): Multiscale and Multiphysics Processes in Geomechanics, SSGG, pp. 1–4.
springerlink.com

Digital Image Correlation (DIC) is a non-destructive method for measuring local surface displacements by mapping pixel gray level values within pixel clusters. The spatial resolution of the resultant local displacement data here is about a sand grain. Note, the resulting displacements do not physically reflect behavior of individual particles. Rather, the data reflect average *incremental* displacements of approximately 7-particle-diameter clusters. Incremental DIC analyses were accumulated over approximately 10% shear strain windows to enable capture of non-instantaneous force chain collapse events. Large strain rotational and volumetric strains were computed locally along and across shear bands from the DIC data.

3 MESO-SCALE STRUCTURES: VORTICES

Figures 1a and b show magnitudes of large strain rotation and Jacobian (volumetric strain), respectively, along the length of shear band for an MC sand specimen at the global softening-critical state transition (Fig. 2). Sense of shear is the top of the shear band moving left relative to the bottom. The pattern in the kinematics is systematic and typifies a multi-force chain collapse event (see [6]). Peaks (absolute value) in rotation indicate regions where force chains are buckling; and, material ahead of these force chains is volumetrically contracting, whereas material behind the buckling force chains is dilating.

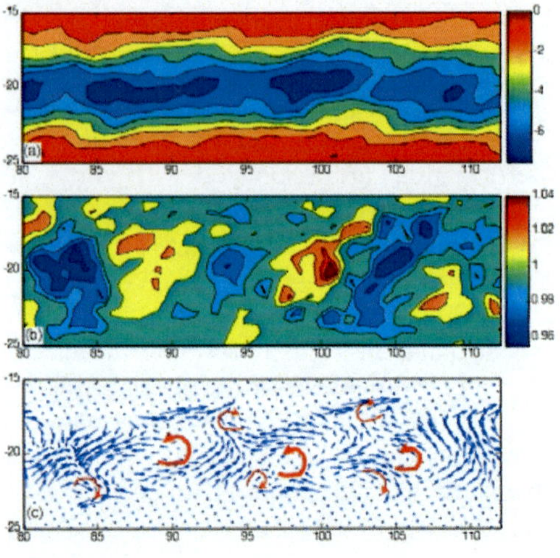

Fig. 1. Kinematic and residual displacements within a shear band in MC sand at the softening-critical state transition. Axes indicate position (in mm) along the shear band: (a) Rotational strain (degrees), (b) Jacobian, (c) residual displacements. In (c), thick arrows highlight vortices and wakes.

To investigate the nature of meso-scale particle movements that manifest the kinematic responses seen in Figs. 1a and b, we first note that in the reference frame of the actual experiment (axial load applied vertically downward), upon shear band formation, the sand above a shear band translates downward with the imposed axial strain rate, and the material below the band translates laterally with the unconstrained "sled". By subtracting from the actual displacement field the base sled movements, we essentially make the base of the shear band stationary. A linearly varying shear displacement field across the shear band was then subtracted from the observed non-affine displacement field, resulting in a "residual" displacement field (Fig. 1c). Distinct, co-rotating, vortices are seen. The vortex centers vary approximately periodically and line up nearly exactly with the peaks (absolute value) in rotational strain (Fig. 1a). At the conflux between adjacent vortices, where residual displacements are in opposition, local volumetric contraction is occurring. Tordesillas et al. [5] saw "microbands" of slip at these junctures, and that we do not perhaps emphasize the strong roles particle interlocking and inter-grain rotational resistance play in a real sand.

A further observation in Fig. 1c is the presence of opposite-trending rotational entities near the shear band boundaries, in between the main shear band vortices. As local maxima in rotational strain do not coincide with these features (Fig. 1a), we believe these may be "wakes" from the main vortices. It is interesting to note that Tordesillas et al. [5] did not observe these features.

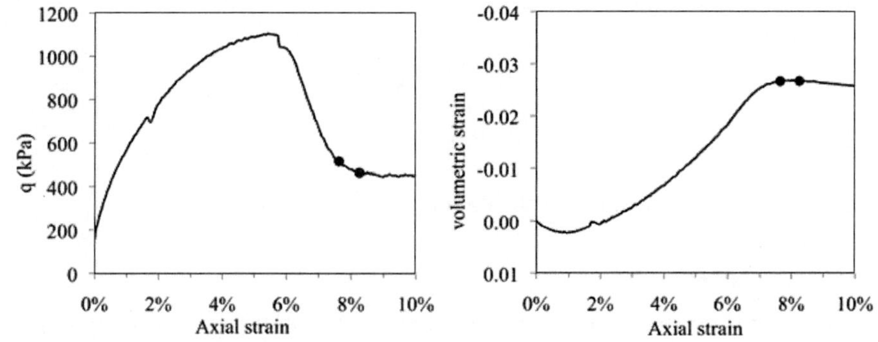

Fig. 2. Global stress-strain behavior for plane strain compression test on dense MC sand. Data points indicate analysis increment of Figure 1.

4 PATTERNS AND EVOLUTION OF VORTICES

Figure 3 shows for an SC sand specimen residual shear band displacements for consecutive 10% shear strain increments at: (a) the softening-critical state transition (i.e. similar to Fig. 2), and (b) the beginning of critical state. Sense of shear is to the right (opposite to Fig. 1). The vortex patterning at the softening-critical state transition (Fig. 3a) is nearly identical to that for the MC sand (Fig. 1c). While the spacing between vortex centers is indeed larger for the larger-grained MC sand, a consistent trend has not been observed for the sands tested here, so more research is needed. New force chains appear to arise at the conflux of opposing displacements between adjacent vortices upon their dissolution.

We use the sequence of residual displacements from Figs. 3a to b to glimpse the fate of our vortex structures. Distinct vortices are not seen in Fig. 3b. However, consistently, the regions between vortices in Fig. 3a are seen to evolve to the "upward" moving regions in Fig. 3b (for example, x-axis positions of -39 and -32). We note that in the residual displacement field, perpendicular and parallel components of displacement both are extremely small (hundredths of mm's), so these "upward" movements are exaggerated. Perpendicular movement in the true displacement field would reflect only subtle undulations in motion: downward movement, for example, may be associated with force chain buckling, and upward movement local dilation or force chain formation. We thus offer that "upward" movements in Fig. 3b indicate formation of new force chains, which are born at the conflux of opposing displacements between vortices.

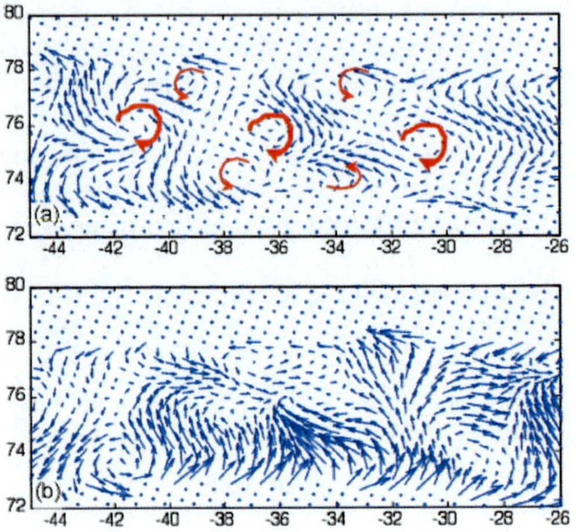

Fig. 3. Residual displacements at: (a) end of softening, (b) beginning of critical state.

5 CONCLUSIONS

Co-rotating vortices are observed within shear bands in sands undergoing plane strain compression. The spatial positioning of these vortices provide meso-scale rationales for observed fluctuations in rotational and volumetric strains along shear bands. Opposite-trending, smaller rotational regions were observed near shear band boundaries between vortices, and may be a byproduct of the high degree of interlocking in real sands. New force chains appear to be born at the conflux of opposing displacements between adjacent vortices.

REFERENCES

[1] Iwashita, K., Oda, M.: Micro-deformation mechanism of shear banding process based on modified distinct element method. Powder. Technol. 109, 192–205 (2000)

[2] Oda, M., Iwashita, K.: Study on couple stress and shear band development in granular media based on numerical simulation analyses. Int. J. Eng. Sci. 38, 1713–1740 (2000)

[3] Rechenmacher, A.L.: Grain-scale processes governing shear band initiation and evolution in sands. J. Mech. Phys. Solids. 54, 22–45 (2006)

[4] Alonso-Marroquin, F., Vardoulakis, I., Hermann, H.J., Weatherley, D., Mora, P.: Effect of rolling on dissipation in fault gouges. Phys. Rev. E. 74, 31306 (2006)

[5] Tordesillas, A., Muthuswamy, M., Walsh, S.D.C.: Mesoscale measures of nonaffine deformation in dense granular assembelies. J. Eng. Mech. 134(12), 1095–1113 (2008)

[6] Rechenmacher, A.L., Abedi, S., Chupin, O.: Evolution of force chains in shear bands in sands, Géotechnique 60(5), 343–351 (2010)

EVOLUTION OF STRAIN LOCALIZATION OBSERVED IN PLANE STRAIN EXPERIMENTS ON CLAY

Dunja Perić[1,*] and Marte Gutierrez[2]

[1] Department of Civil Engineering
Kansas State University
2118 Fiedler Hall
Manhattan, Kansas 66506-5000
* e-mail: peric@ksu.edu, Web page: http://www.ksu.edu/facultystaff.peric
[2] Division of Engineering
Colorado School of Mines
1500 Illinois Street
Golden, Colorado 80401
e-mail: mgutierr@mines.edu, web page: http://www.mines.edu/~mgutierrez

Summary. One of the simplest forms of strain localization takes place in a biaxial or plane strain apparatus, in which a homogeneous sample gradually becomes fragmented usually into two nearly rigidly deforming bodies that are interconnected by an extensively deforming shear band. Although clayey soils are largely encountered in practice, a vast majority of experimental studies conducted in plane strain devices has been devoted to sands, whereby the grain scale processes have been successfully quantified to a large extent. The major goal of this study was to advance understanding of strain localization in clays. A special emphasis herein is placed on quantifying the strain localization at sub-macroscopic scales.

Keywords: clay, meso-scale, plane strain, strain localization.

1 INTRODUCTION

Strain localization is a ubiquitous feature of elastic-plastic materials including soils. It is characterized by a sudden appearance of narrow zones, within which large strains evolve with continued loading. Little or no strain occurs outside of these zones, which are also known as deformation bands. The onset and subsequent evolution of strain localization signifies the transition from a relatively uniformly deforming amorphous soil mass to the increasingly structured medium. The internal structure comprises a number of nearly rigidly moving bodies that are bounded by deformation bands, thus ultimately resulting in a fragmented soil mass. It is because the soils around foundations, in slopes, excavations and embankments, earth dams and around tunnels experience such failure mechanisms under the conditions that are conducive to strain localization that this subject is of an enormous importance for the advancement of engineering practice.

R.I. Borja (Ed.): Multiscale and Multiphysics Processes in Geomechanics, SSGG, pp. 5–8.
springerlink.com

2 EXPERIMENTAL PROGRAM

A series of undrained plane strain compression experiments were conducted on the one dimensionally, slurry consolidated kaolin clay. The liquid and plastic limits of the clay are 52% and 27% respectively while its specific gravity is 2.66. The 100% of particles by weight have an average diameter that is smaller than 0.075 mm, and the mean particle size is about 0.00084 mm according to the results of the hydrometer test. The 140 mm x 40 mm x 80 mm prismatic clay samples were placed between lubricated and glass lined bottom and side platens, and subjected to a one-dimensional consolidation followed by an undrained compression to failure. The bottom platen rests on a linear bearing or sled, which facilitates an uninhibited onset and evolution of strain localization. The device is internally instrumented including seven internal linear variable differential transducers (LVDT), out of which one pair was used to measure the axial displacement, the additional two pairs to measure the horizontal displacements near the top and bottom of a clay sample, and the last one to measure the sled displacement. More detailed description of the biaxial apparatus used in this study was provided by Finno and Rhee [1].

3 MESO-SCALE EVOLUTION OF STRAIN LOCALIZATION

Multiple boundary displacement measurements have been used to characterize the shear band development in soft rock [2] and sand [3]. The essentially different particle sizes and shapes of sands and clays are bound to have significant effects on the grain scale processes, which are vitally important for strain localization due to the small thicknesses of shear bands. While the thickness of shear band in sands is about 10–20 times larger than the mean particle size [4], the corresponding information for clays is scarce at best. According to [5], the zone of particle reorientation in kaolin and bentonite clays ranged from 2 to 7 mm. The same study suggested that the thickness of shear band was of the order of 1mm, based on the density measurements by X-ray scanner.

The characterization of grain scale processes in clays presently remains elusive due to their extremely small particle sizes, which are three orders of magnitude smaller than sand particles. A shear band is a sub-macro-scale structure because: 1) its thickness is at least one order of magnitude smaller than its width and length, the later being of the same order of magnitude as the entire biaxial sample, and 2) it interconnects the macro-scale to the sub-macro scales. Herein, characterization of the strain localization evolution is conducted based on the multiple boundary measurements by employing the following assumptions: 1) the shear band is inextensible, it deforms in a simple shear mode, 2) displacements of the shear band vary in a linear fashion along its thickness, but are constant along its length, thus rendering its deformation uniform, 3) after the shear band is fully formed the macro-scale axial and lateral displacements are nearly fully absorbed within the shear band, the outside material is either deforming as a rigid body or unloading elastically, and 4) the orientation of the shear band is fixed in time. The measured lateral displacements of the portion of the biaxial sample that was not intersected by the shear band were extremely small resulting in a 0.15% of a possible decrease of the length of the shear band, thus confirming the assumption of the inextensible shear band. The orientation of shear band was found to vary about 1.9 degrees total with the progress of strain localization, but this change was neglected for the purpose of these preliminary results. The average orientation was calculated and used instead.

Figure 1 depicts the evolution of the total dilatancy angle calculated from the normal and slip displacements, which were inferred from boundary measurements and the average shear band orientation. A positive dilatancy angle indicates an increase in the thickness of the shear band. In accordance with [2], the elastic unloading displacement is deducted from the vertical displacement measured at the boundary to obtain the vertical displacement of the shear band. Four different assumptions regarding the elastic unloading are included to reflect the possible range of elastic properties that affect the amount of unloading. While the unloading in the horizontal direction is negligible as noted above, the unloading in the vertical direction is sufficient to affect the dilatancy angle. This is the consequence of both, the consolidation- induced cross-anisotropy and the sample having larger height than the width.

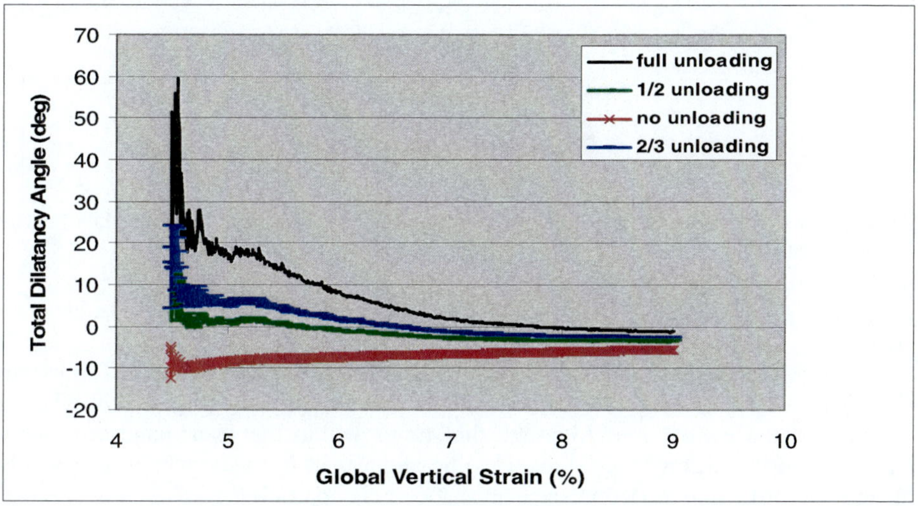

Fig. 1. Total dilatancy angle of the shear band for a test on slightly heavily overconsoliated clay

The full formation of the shear band is defined as a stage, at which the shear band intersects the entire sample. At this point the sled started to move at a constant velocity, which was closely followed by a decrease of the excess pore water pressure. It is because the excess pore water pressure continued to decrease until about 6.4% of the global vertical strain that the assumption of two-third of a full unloading is adopted for calculation of strains. Namely, at a 6.4% of the global vertical strain the excess pore water pressure starts to rise slightly, which corresponds to a change of sign of a dilatancy angle shown in Figure 1.

The Green-Lagrange in-plane shear and volumetric strains inside the shear band are calculated based on its displacements by assuming two different initial thicknesses of the shear band as shown in Figure 2.

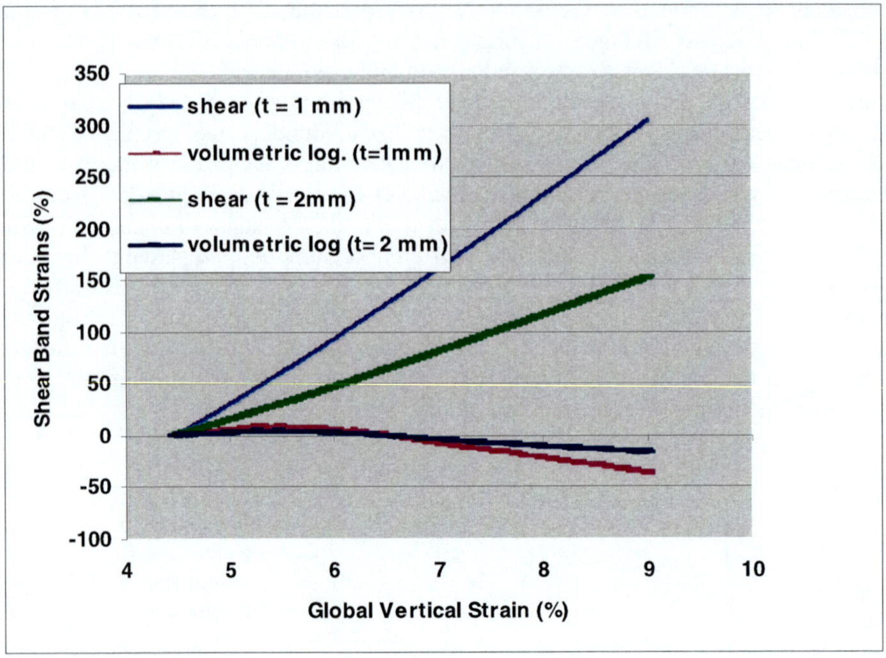

Fig. 2. Shear band strains for two different shear band thicknesses for a test on slightly heavily overconsolidated clay

4 CONCLUSIONS

Local displacements and strains show that the shear band, which formed in slightly heavily overconsolidated clay sample was initially dilating, followed by a transition to the contraction. Although the strains within the shear band are highly dependent on its thickness, they appear to be larger than in sands, which is plausible due to the significantly different size and shape of clay particles.

REFERENCES

[1] Finno, R.J., Rhee, Y.: Consolidation, pre- and post-peak shearing responses from internally instrumented biaxial compression device. Geotechnical Testing Journal 16, 496–509 (1993)

[2] Labuz, J.F., Dai, S.T.: Residual strength and fracture energy from plane-strain testing. J. of Geotechnical and Geoenvironmental Engineering 126, 882–889 (2000)

[3] Han, C., Vardoulakis, I.G.: Plane-strain compression experiments on water-saturated fine-grained sand. Géotechnique 41, 49–78 (1991)

[4] Mooney, M., Finno, R.J., Viggiani, G.: A unique critical state for sand? J. of Geotechnical and Geoenvironmental Engineering 124, 1100–1108 (1998)

[5] Hicher, P.Y., Wahyudi, H., Tessier, D.: Microstructural analysis of strain localization in clay. Computers and Geotechnics 16, 205–222 (1994)

LARGE SCALE SHEAR BANDING IN EXTENSION

Hans Mühlhaus[1,*], Louise Olsen-Kettle[1], Jingyu Shi[1], and Louis Moresi[2]

[1] ESSCC, The University of Queensland, Brisbane, 4072, Australia
* e-mail: h.muhlhaus@uq.edu.au, l.kettle@uq.edu.au, jingyushi88@hotmail.com
[2] School of Geosciences, School of Mathematical Sciences, Monash University, Clayton, Victoria, 3800, Australia
e-mail: Louis.Moresi@sci.monash.edu.au

Summary. In this paper, we will explore the role of non-coaxiality on shear banding in pure shear. We first outline the consitituve relations. The deformation and localization process is illustrated by results of large deformation finite element simulations on a rectangular domain in extension for different constitutive models. We also show the variation of the average horizontal stress (stress resultant) conjugate to the prescribed boundary velocity with a strain measure for the horizontal extension. We assume incompressible deformations since the emphasis in this study is on large deformations. The equations of motion are integrated using an updated Lagrangian scheme.

Keywords: mesh sensitivity, non-coaxial flow, shear banding.

1 INTRODUCTION

Shear bands or faults are arguably the most important deformation mechanisms in the upper lithosphere. They occur on many different scales, e.g. in the form of detachment faults in rift zones or as collapse mechanisms of geotechnical structures. The conditions under which shear bands form depend strongly on the local material behavior and, as such, on the details of the constitutive relationships. If the flow rule is non-associated, e.g. if the pressure sensitivity is not equal to the dilatancy factor relating the inelastic volume change to the equivalent plastic strain rate, then shear banding is possible under increasing stress, in the hardening regime. The stress level and the preferred band orientation also depend on whether the plastic strain rate has the same principal axes as the stress. If the latter is the case then the flow rule is designated as co-axial. Non-coaxial models have been proposed originally by de Jong [1,2] and Spencer [3,4] in the context of granular materials. A number of researchers have considered the theory, looking at new derivations, various motivations and considering extensions [5-7]. See Tordesillas *et al* [8] for a recent review.

2 CONSTITUTIVE RELATIONSHIPS

We derive a simple basic plane-strain non-coaxial viscous-plastic model by assuming that the plastic deformation is carried by a single slip system (Figure 1) which is oriented under either $(\pi/4+\nu/2)$, i.e. $(\mathbf{s}^a,\mathbf{n}^a)$ or $-(\pi/4+\nu/2)$, i.e. $(\mathbf{s}^b,\mathbf{n}^b)$ to the direction of the largest (least compressive) principal stress. The angle ν is the angle of non-coaxiality (de Jong [1]; Spencer [4]), which is the angle between

R.I. Borja (Ed.): Multiscale and Multiphysics Processes in Geomechanics, SSGG, pp. 9–12.
springerlink.com

the direction of larger principal stress and the direction of larger principal strain rate. The flow rule is coaxial if $\nu = 0$ and non-coaxial if $\nu > 0$. The flow rules proposed by Moresi and Mühlhaus [9] are obtained for $\nu = \phi$, where ϕ is the Mohr-Coulomb angle of friction. The following relationship exists between the plastic stretching D_{ij}^P, the shear and normal vectors (s_i and n_j, respectively) in the drawing plane, and the plastic shear strain rate $\dot{\gamma}^P$:

$$D_{ij}^P = \frac{\dot{\gamma}^P}{2}(s_i n_j + s_j n_i). \tag{1}$$

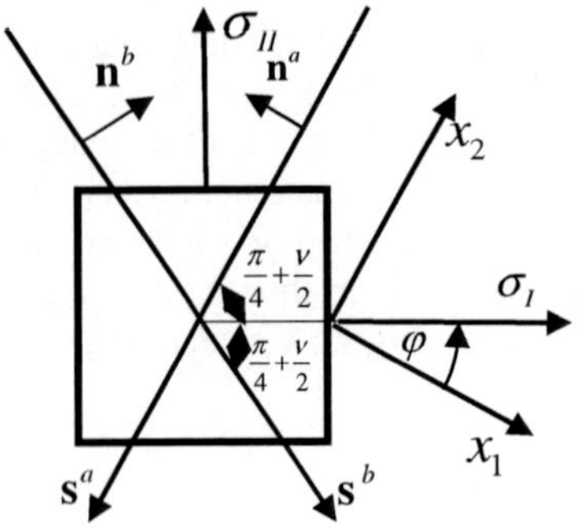

Fig. 1. Principal stresses (σ_I, σ_{II}) and slide systems $(\mathbf{s}^a, \mathbf{n}^a)$ and $(\mathbf{s}^b, \mathbf{n}^b)$; (x_1, x_2) are the coordinates of a global, spatially fixed Cartesian coordinate system.

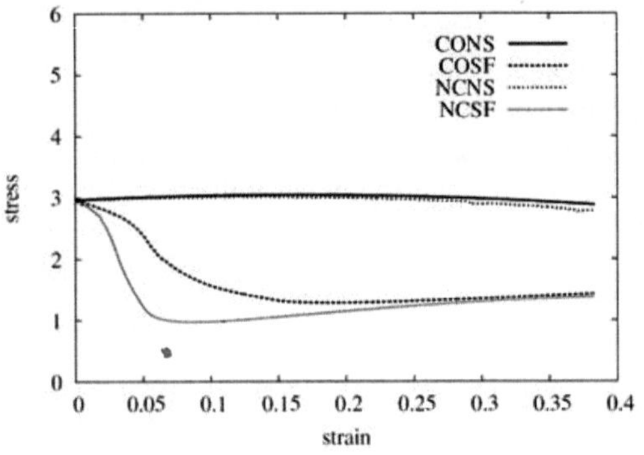

Fig. 2. Variation of average horizontal stress with horizontal strain extension

The components of the shear and the normal vectors in (1) depend on the orientation of the shear mechanism relative to the principal stress axes. The components of the shear and normal vector can be expressed in terms of trigonometric functions of $\varphi \pm (\pi/4 + \nu/2)$, where φ is the angle between the spatial coordinate system and the axis of the larger principal stress (Figure 1). Using the well-known relationships for the principal stress angle φ and the stress components, we obtain for the plastic stretching:

$$D_{ij}^P = \frac{\dot{\gamma}^P}{2\tau}(\cos\nu\sigma'_{ij} \pm \sin\nu\sigma_{ij}^{\perp}) \tag{2}$$

where the + sign applies if $(\pi/4 + \nu/2)$ is active, and the – sign applies if $-(\pi/4 + \nu/2)$ is active. In this relation, $\sigma_{ij} = \sigma'_{ij} - p\delta_{ij}$, with $p = -(\sigma_{11} + \sigma_{22})/2$ and δ_{ij} being the Kronecker delta tensor and we have the relationships:

$$\boldsymbol{\sigma}^{\perp} = \mathbf{Q}_{\theta=\pi/4}\boldsymbol{\sigma}'\mathbf{Q}_{\theta=\pi/4}^T, \quad \mathbf{Q} = \begin{bmatrix} \cos\theta & -\sin\theta \\ \sin\theta & \cos\theta \end{bmatrix}, \quad \tau = \sqrt{\frac{1}{2}\sigma'_{ij}\sigma'_{ij}} = \sqrt{\frac{1}{2}\sigma'^{\perp}_{ij}\sigma'^{\perp}_{ij}}. \tag{3}$$

It follows that $\dot{\gamma}^P = \sqrt{2D_{ij}^P D_{ij}^P}$. We assume that plastic flow takes place if the Mohr-Coulomb yield criterion is satisfied. In this case, we have $\dot{\gamma}^P > 0$ if $\tau = c\cos\phi + p\sin\phi$ and $\dot{\gamma}^P = 0$ if $\tau < c\cos\phi + p\sin\phi$, where c is the cohesion.

We follow the standard approach by decomposing the total strain rate into a viscous and a plastic part, i.e. $D_{ij} = D_{ij}^v + D_{ij}^p$. We assume the viscous part of the stretching is given by $D_{ij}^v = \sigma'_{ij} / 2\eta$ and define plastic viscosity as $\eta^p = (c\cos\phi + p\sin\phi) / \dot{\gamma}^p$. It then follows:

$$\mathbf{D} = \frac{1}{2}\left(\frac{1}{\eta\cos\nu} + \frac{1}{\eta^p}\right)\cos\nu\sigma'_{ij} \pm \frac{1}{2\eta^p}\sin\nu\sigma_{ij}^{\perp}. \tag{4}$$

The effect of material damage accumulation, e.g. through micro-crack growth or porosity nucleation is considered by the simple strain softening relationship proposed by Moresi and Mühlhaus [9]: $c = c_0 + (c_\infty - c_0)\min(1, \gamma^p / \gamma_\infty)$, where c_0 is the initial cohesion, c_∞ is the residual cohesion, γ^p is the accumulated plastic strain and the parameter γ_∞ is the value of the accumulated plastic strain at which the cohesion assumes its residual value.

3 EXTENSION OF A RECTANGULAR LAYER

We conduct numerical simulations on a rectangular domain with initial dimensions $(L_0, H_0)=(3,1)$ under extension. Gravity is an external force acting in the domain. The simulation is carried with non-dimensionalized governing equations, with the density of the material $\rho = 1$, and all parameters are non-dimensional. All boundaries are assumed to be smooth. The top boundary is assumed as stress free. The vertical velocity is assumed to be zero at the bottom boundary. The velocities $V_1 = 0.035$ and $V_1 = -0.035$ are prescribed at the right and left boundaries, respectively. We also assume a small weak zone (0.04×0.04) of constant viscosity $\eta_w = \eta / 100$ located at the middle of the bottom boundary, where $\eta = 100$ is the viscosity of the surrounding material. We choose the internal friction $\tan\phi = 0.4$, the initial cohesion $c_0 = 4.0$, the residual cohesion $c_\infty = 2.0$ and $\gamma_\infty = 0.1$.

Figure 2 shows the variation of the average normal stress on the right boundary with the horizontal strain, $\varepsilon = \ln(L / L_0)$, for various material models. The coaxial (CO) model results from an angle of non-coxiality of $\nu = 0$ and non-coaxial (NC) model has an angle of non-coaxiality of $\nu = \phi$. The cohesion $c = c_0$ gives the non-softening (NS) case. The softening (SF) effect is governed by the softening relationship given earlier. If there is no softening, then non-coaxiality has very little effect. For softening materials at large strain, the effect of non-coaxiality disappears.

Figure 3 shows the distribution of the second invariant of strain rate with deformed configuration for various models under extension at strain level $\varepsilon = 0.1$. For coaxial materials without softening, there are less distinguishable bands emerging from the weak zone where the second invariant of the strain rate in the bands does not differ much from that in the remaining region (see Figure 3a). The bands disappear at a later stage. If the material softens, then there is no clear picture with two bands. Instead, multiple bands emerge and develop and a strain concentrating region is found at the opposite site of the weak zone (see Figure 3b). At later stages, two bands emerge from the region and join the neighboring bands. For non-coaxial materials, three strain concentrating regions start on the top surface. The two off-centre surface concentrations correspond to outcrops of shear bands nucleated from the bottom centre perturbation and the centre surface concentration represents an outcrop of a tensile-separation (Mode I) fracture. If the material does not soften, extension of the bands terminates and more short bands emerge from the top surface (Figure 3c). If the material softens, then the bands from the two off-centre regions continue to extend and reach the weak zone (Figure 3d). This pattern remains until end of the simulation.

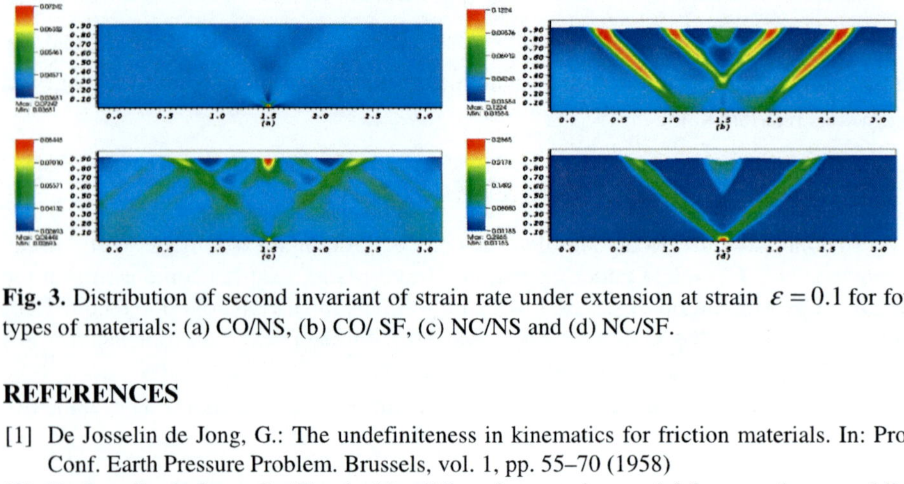

Fig. 3. Distribution of second invariant of strain rate under extension at strain $\varepsilon = 0.1$ for four types of materials: (a) CO/NS, (b) CO/ SF, (c) NC/NS and (d) NC/SF.

REFERENCES

[1] De Josselin de Jong, G.: The undefiniteness in kinematics for friction materials. In: Proc. Conf. Earth Pressure Problem. Brussels, vol. 1, pp. 55–70 (1958)

[2] De Josselin de Jong, G.: The double sliding, free rotating model for granular assemblies. Geotechnique 21, 155–162 (1971)

[3] Spencer, A.: A theory of the kinematics of ideal soils under plane strain conditions. J. Mech. Phys. Solids 12, 337–351 (1964)

[4] Spencer, A.J.M.: Deformation of ideal granular materials, mechanics of solids: the Rodney Hill Anniversary Volume, pp. 607–652. Pergamon Press, Oxford (1982)

[5] Mehrabadi, M.M., Cowin, S.C.: On the double sliding free rotating model for the deformation of granular materials. J. Mech. Phys. Solids 29, 269–282 (1981)

[6] Harris, D.: A unified formulation for plasticity models of granularand other materials. Proc. R. Soc. Lond. A. 450, 37–49 (1995)

[7] Jiang, M.J., Harris, D., Yu, H.S.: Kinematic models for non-coaxial granular materials, Part II. Int. J. Numer. Anal. Meth. Geomech. 29, 663–689 (2005)

[8] Tordesillas, A., Shi, J., Muhlhaus, H.: Non-coaxiality and force chain evolution. Int. J. Engng Sc. 47, 1386–1404 (2009)

[9] Moresi, L., Mühlhaus, H.: Anisotropic viscous models of large-deformation Mohr-Coulomb failure. Philosophical Magazine 86, 3287–3305 (2006)

SHEAR BANDING IN GEOMATERIALS IN PLANE STRAIN: PHYSICAL AND ANALYTICAL MODELS

Theodoros Triantafyllidis[1], Lars Röchter[2], and Andrzej Niemunis[1]

[1] KIT Karlsruhe Institute of Technology
Institute for Soil Mechanics and Rock Mechanics
Engler-Bunte-Ring 14, 76131 Karlsruhe
e-mail: Triantafyllidis@kit.edu, web page: http://www.ibf.uni-karlsruhe.de
[2] Ruhr University Bochum
Chair of Foundation Engineering, Soil and Rock Mechanics
Universitätsstraße 150, 44780 Bochum
e-mail: lars.roechter@rub.de, web page: http://www.ruhr-uni-bochum.de

Summary: The proposed method for the determination of the distance between shear bands under plane strain extension is based on the assumption that the material is softening inside the shear bands and is undergoing elastic unloading outside. The observed geometric pattern corresponds to derivation based on the principle of the lowest potential energy rate.

Keywords: geomaterial, modeling, plane strain, shear banding.

1 INTRODUCTION

Shear banding can be detected in physical model tests as well as in engineering scale problems, e.g. behind retaining walls and in the geological or tectonic scale. A literature review is given, e.g. by Wolf et al [1] and Röchter et al [3] especially for parallel normal faults in situ and also for extensional model tests and the respective numerical studies. During extensional loading in sand-box experiments, Wolf [2] and Röchter [4] have paid special attention to the quantitative determination of the shear band spacing by comparing the results of physical and analytical models.

Analysing sand-box model experiments, Wolf [2] reported a simultaneous onset of parallel shear bands, like in a bookshelf mechanism in an extensional device, whereby the movement of the blocks between the shear bands is a rotation towards the stretching direction. Based on the dissipated energy of the whole system and considering shearing, dilation and softening inside the shear band and elastic unloading outside the shear band, Niemunis [5] calculated the width of the blocks by maximization of the dissipated energy. The spacing between the shear bands turns out to be smaller when the softening is more pronounced. This confirms findings from experimental results with dense and loose sand [2].

Mandl [6] also referred to parallel shear bands in sandbox extension experiments, in which the basal layer is stretched. After the localization of deformation a decrease of shear stress from the limit state to smaller stress state took place. Between the shear bands the material blocks were elastically unloaded with the horizontal stress $\sigma_h = \sigma_3$ (minor stress) increasing, Fig. 1. Mandl considered the strain softening as a kind of brittleness parameter being the most important property in determining the spacing of parallel faults. According to the model in materials with steeper softening (i.e. or with

R.I. Borja (Ed.): Multiscale and Multiphysics Processes in Geomechanics, SSGG, pp. 13–16.
springerlink.com

more brittle behaviour), the released elastic energy is larger leading to a wider spacing. For constant strain rate at the bottom of the sand box the elastic modulus for reloading and the strain softening are the competing parameters.

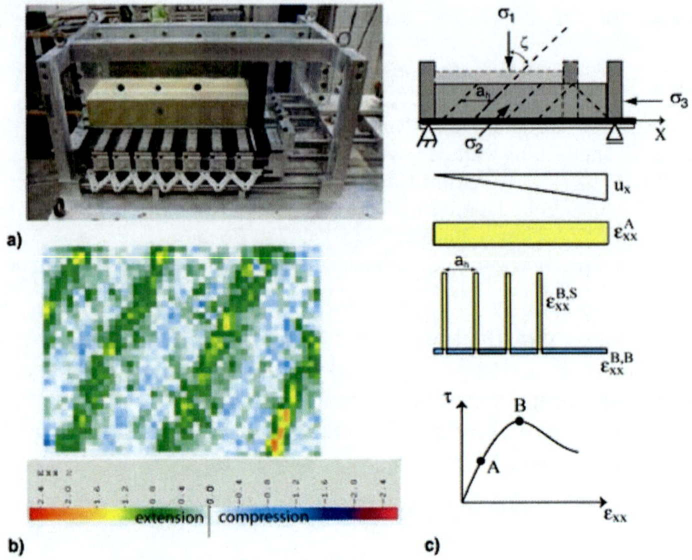

Fig. 1. a) Sand box extensional device, b) DIC resolution with localization and unloading zones c) schematic representation of the strains before and after localization.

Wolf [2] and Röchter [4] performed numerous model tests in an extensional device in order to study shear banding using the DIC (digital image correlation) method. Different materials (cohesion-frictional) with different densities and granulometry were tested under normal and increased gravity in a centrifuge. In Fig. 1, a photograph of the device and the schematic sketch of the loading as well as DIC results for the shear bands and the unloaded zones are shown. The main findings are: 1) the shear bands appear in sandy material spontaneously over the specimen's height with the exception of moist clayey material mixtures, 2) the shear band spacing is proportional to specimen's height, 3) the variation of the stress level up to 50g influences neither the spacing between shear bands nor the inclination of the shear bands at the onset of the localization, 4) for sandy material no dependence between deformation velocity as boundary condition and shear band spacing could be detected, 5) a linear dependence between the relative density I_d and shear band spacing could be detected: decreasing density corresponds to larger shear band spacing, 6) the spacing between adjacent parallel shear bands increases with the addition of clay fraction or due to capillary action (apparent cohesion), 7) the spacing of shear bands in dry sand represents the lower bound.

2 ANALYTICAL APPROACH

From the variety of shear banding mechanisms we concentrate only on the so-called "bookshelf mechanism" in this study. For loading processes beyond the peak the solution of the BVP need not be unique. On the basis of an idea of Niemunis [5], it can be assumed that under all possible deformation patterns the material uses the one that leads to the maximum dissipation rate, Fig. 2. The material tries to reach the smallest possible potential energy level in the unloaded zone choosing a respective width of the unloading blocks.

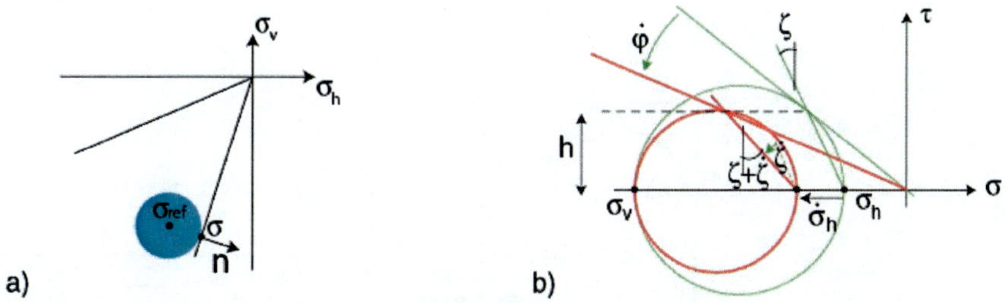

Fig. 2. a) Elastic unloading at a K_a stress condition towards the centre of the elastic locus and b) the anisotropic softening allowing the Mohr circle to protrude beyond the Coulomb surface.

For the description of the elastic energy potential Π for a unit soil volume we cannot use the stress-free configuration because it is absent in soils. At the formation of shear banding it is known that the material reaches the peak friction angle and the material stress state resembles a K_a-condition (Fig. 2) behind a retaining wall. For the unloading a change of the elastic potential can be seen in relation to this state, whereby the centre of the elastic locus is close to the yield condition (Fig. 2a). Therefore, the rate $\dot{\Pi}$ is considered and the material deforms in such way that the width of the unloaded blocks corresponds to the maximal decrease ($\dot{\Pi} < 0$). For simplicity reasons in the analysis, we may assume that the change of the vertical stress due to horizontal stretching is negligible. For the elastic potential changes with reference to the active stress state only the minor stress is considered ($\sigma_h = \sigma_{min}$) and the intermediate stress is regarded as constant. Therefore, only the changes of σ_h are taken into account. The softening may be isotropic or anisotropic (Fig. 2b). An extended approach includes also in the analysis the friction between the soil and the side walls and on the base of the stretching leading to further reduction of the average decrease of the horizontal stress rate $\dot{\sigma}_h$ over the cross-sectional area of the specimen [4]. Using the softening modulus k (k: reduction of friction angle over shear strain increment) and an assumed constant elastic unloading modulus E of the material, the rate of the specific potential energy $\dot{\Pi}_s$ can be expressed analytically as a function of spacing a_h. The minimum of $\dot{\Pi}_s$ over a_h is given as [4]:

$$\dot{\Pi}_s = -\frac{1}{E}\sigma_h\dot{\sigma}_h \Rightarrow \frac{\partial\dot{\Pi}_s}{\partial a_h} = 0 \Rightarrow a_h. \tag{1}$$

The essential input material parameters required in the model are the decrease of the angle of friction versus the shear strain and the elastic unloading modulus in the softening regime. In order to obtain these parameters, Röchter [3, 4] has carried out biaxial plain strain experiments under similar conditions as in the model extension tests (see Fig. 3a). Using local measurements with contact free displacement sensors (CFDS) it was possible to calculate the horizontal deformation inside the shear band after the onset of localization. Taking into account the different shear strain rates within the entire specimen $\dot{\gamma}$, inside the shear band $\dot{\gamma}_B$ and the additional local measurements, Röchter et al [3] derived a relationship between the strain softening inside the shear band k_B and that of the entire specimen k ($\Delta\gamma_B/\Delta\gamma = k/k_B$).

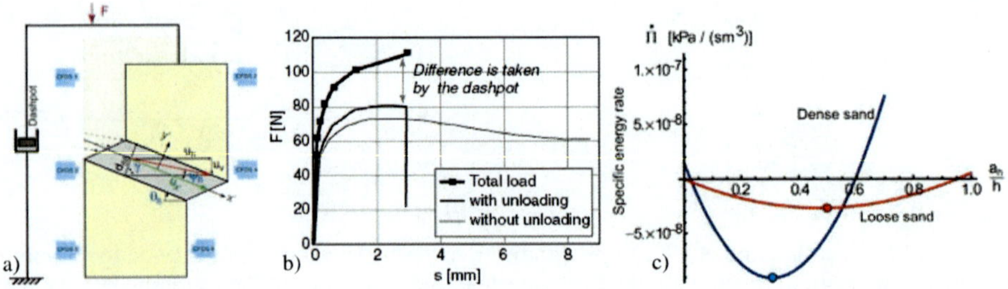

Fig. 3. a) Artificial stress reduction during softening in biaxial test, b) comparison between tests on dense sand with and without unloading while softening and c) determination of the shear band distance a_H using min($\dot{\Pi}$).

For the determination of the unloading modulus in the softening regime, Röchter [4] used a special stress control method. A dashpot allows for an artificial stress reduction during softening in triaxial compression tests. The working principle is sketched in Fig. 3a. During the softening the vertical load is gradually decreased leading to either continuing softening or elastic unloading. The results of such an experiment are presented with and without unloading in Fig. 3b and the analytical results for the spacing are given in Fig. 3c for dense and loose sands. The observations of Wolf [2] regarding the influence of the density on the distance between shear bands were verified.

REFERENCES

[1] Wolf, H., König, D., Triantafyllidis, T.: Experimental investigation of shear band patterns in granular materials. Journal of Structural Geology 25, 1229–1240 (2003)

[2] Wolf, H.: Zur Scherfugenbänderung granularer Materialien unter Extensionsbeanspruchung. In: Triantafyllidis, T. (ed.): PhD thesis, Ruhr-Universität Bochum, vol. 37 (2005)

[3] Röchter, L., König, D., Schanz, T., Triantafyllidis, T.: Shear banding and strain softening in plane strain extension: Physical modelling. Granular Matter 12(3), 287–301 (2010)

[4] Röchter, L.: Systeme paralleler Scherbänder in triaxialer Extension. PhD thesis, Ruhr-University Bochum (2010) (in preparation)

[5] Niemunis, A.: Internal Report (2005)

[6] Mandl, G.: Faulting in Brittle Rocks. Springer, Heidelberg (2000)

MICROMECHANICAL STUDY OF STRAIN LOCALIZATION AND FAILURE IN THE DIRECT SHEAR TEST OF GRANULAR SOILS

Marte Gutierrez[1,*] and Jianfeng Wang[2]

[1] Division of Engineeering
Colorado School of Mines
1610 Illinois St., Golden, CO 80401, USA
* e-mail: mgutierr@mines.edu, web page: http://www.engineering.mines.edu
[2] Department of Building Construction
City University of Hong Kong
Tat Chee Avenue, Kowloon, Hong Kong
e-mail: jefwang@cityu.edu.hk, web page: http://bccw.cityu.edu.hk

Summary. This paper presents a study of the micromechanical response of granular materials in the Direct Shear Test (DST) using DEM (Discrete Element Method). Parametric studies were conducted to investigate the effects of specimen length and height (in relation to the particle size) on the bulk granular material shear strength and shear banding behavior in the DST. A mesh-free strain calculation method previously developed by the authors was used to capture and visualize the evolution of strain localization and failure inside the direct shear box.

Keywords: direct shear test, discrete element method, progressive failure, specimen size scale effects.

1 INTRODUCTION

Specimen scale effects on the shear strength parameters of granular soils obtained from the Direct Shear Test (DST) has been a well-studied topic since 1930's. The nature of the problem lies in the artificial boundary effects introduced by the shear box on the deformation and strength behavior of soils. When the shear box apparatus does not correctly simulate the actual field conditions under which the soil deforms, doubt is cast upon the reliability of test results to be used in practical engineering design. In view of the popularity of the DST, due to its simplicity and lower testing cost compared with other laboratory tests, this paper presents a detailed study of the DST scale effects using the DEM. Two-dimensional DEM simulations of the direct shear box test filled with rigid spherical particles of uniform or varying sizes are performed using PFC2D (Particle Flow Code 2D) developed by Itasca [1]. Different variables are considered in a parametric study so that the specimen scale effects can be captured for different specimen configurations. Macroscopic model behavior is interpreted using stress and volume change data of the bulk material measured at the boundaries. Furthermore, using the strain calculation method previously developed by the authors [2], the localized shear zone developed inside the specimen as a function of the above variables is visualized and analyzed.

R.I. Borja (Ed.): Multiscale and Multiphysics Processes in Geomechanics, SSGG, pp. 17–20.
springerlink.com

2 DEM SIMULATIONS

The DEM model of the DST is created in an effort to closely mimic the physical direct shear apparatus in the laboratory. Assemblages of rigid spherical particles with varying sizes are generated to fill the specified volume and allowed to consolidate at a particular interparticle friction coefficient to achieve a target initial porosity. For all the granular specimens used, the median particle diameter (D_{50}) is equal to 0.7 mm. The largest size of the shear box used in the current study is 88 mm (L) × 56 mm (H), with both the upper and lower half of the box equal to 28 mm. This size gives the largest ratios of box length and box height to median particle diameter of about L/D_{50} = 126 and H/D_{50} = 80. Simulation results from these dimensions have been validated against laboratory direct shear test results using 0.7 mm glass beads by [3]. Glass bead properties measured in the laboratory, such as particle density, shear modulus and interparticle friction coefficient were directly used in the simulations discussed herein. Specimens are first consolidated to equilibrium under a specified vertical stress applied on the top boundary through a servo-control mechanism. A uniform interparticle friction coefficient of 0.3 is used in the initial consolidation stage to minimize the variation of initial porosity due to particle interlocking. A high value of 0.9 is selected for the particle to wall friction coefficient to prevent any slip between the bulk material and the boundary walls. In addition, the top and bottom boundaries were further roughened by introducing continuous 1 mm wide sawtooth-shaped grooves as recommended by the ASTM Standard [4].

3 RESULTS AND DISCUSSIONS

The results of the DEM simulations indicated that the peak stress ratio increases with decreasing box length regardless of the normal stress level and specimen gradation. However, an opposite trend is found between the peak stress ratio and box height, with a few minor exceptions. The above macroscopic effects can be better demonstrated when they are correlated to the box aspect ratio (L/H), as shown in Fig. 1. In this figure, the trends of peak stress ratio against the aspect ratio are clearly marked by two groups of arrows that suggest the variations of peak stress ratio against the box length scale and height scale, respectively. Evidently, the peak stress ratio decreases with increasing aspect ratio in both directions but a higher rate of change is found in the direction of constant height scale and decreasing length scale. This observation suggests that the box length scale may play a more dominant role in controlling the shear behavior of granular specimen than the box height scale. The numbers in Fig. 1 beside each data point indicate the applied shear strains at each corresponding peak stress state. We can see that this value increases with decreasing L/H when either L or H is fixed.

It can be inferred from the above observations that the effects of progressive failure become more pronounced with the increase of box aspect ratio (L/H), which could happen in either of the following two cases: increasing box length and a fixed box height, or a fixed box length and decreasing box height. Apparently, for the large aspect ratio, the materials in the middle of the box length experience less shear straining than those closer to the lateral boundary in the small strain stage, meaning that the shear strength of the middle portion cannot be fully mobilized at the first peak stress state. When *L/H* becomes smaller, a larger percentage of the specimen including more material near the middle of the box tends to be involved in intense shear at the peak state, implying a more global failure of the whole specimen.

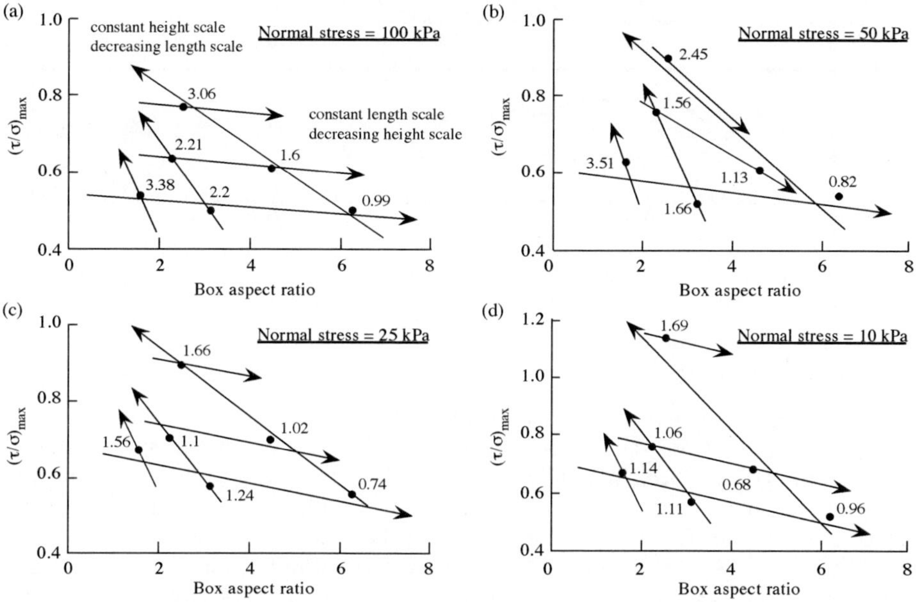

Fig. 1. Effects of the box aspect ratio (*L/H*) on the peak stress ratio.

The effect of local/global failure at the peak stress state and the ensuing progressive failure as affected by the box length scale can be better visualized in terms of shear strain localization calculated by the method previously developed by the authors [2]. Figure 2 shows the shear strain distributions of the three simulations at their peak stress states. Clearly, a more pronounced and concentrated shear band in the mid-plane of the box is found in the 35 mm case (Fig. 2a). In comparison, less developed shear bands deviating from the mid-planes of the box are found in the 63 mm and 88 mm cases (Figs. 2b and 2c). In the latter two cases, rather than global and well-defined, shear bands are more local and less pronounced. Indeed, the bands are mainly composed of local structures subordinate to the primary structure of the shear band along the mid-plane during the subsequent shearing process [5,6].

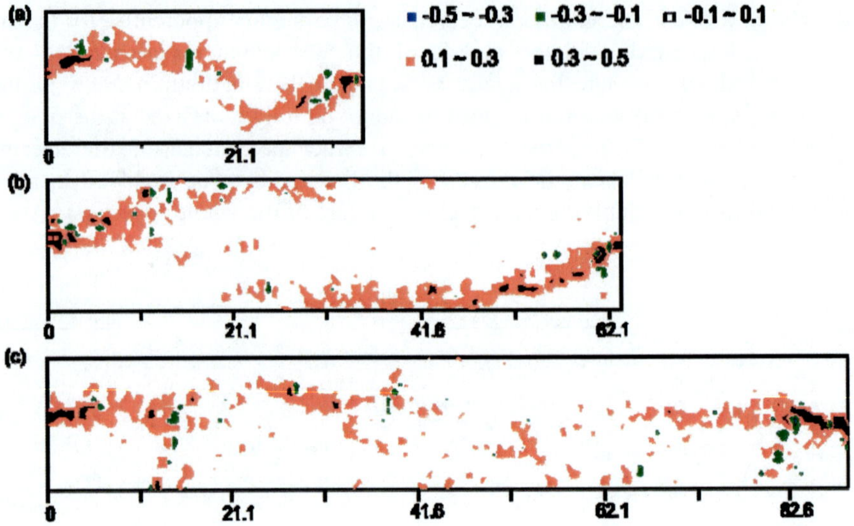

Fig. 2. Shear strain distributions inside the shear boxes of the first group of simulations captured at the peak stress state: (a) $L = 35$ mm; (b) $L = 63$ mm; (c) $L = 88$ mm.

4 CONCLUSIONS

Micromechanics-based analysis showed that the local and global aspects of fabric change and failure are the major mechanisms responsible for the specimen scale effect in the Direct Shear Test. Global failure along the primary shear band prevails when the specimen length scale, and the height to length aspect ratio are small, while progressive failure becomes more likely when the specimen length scale and aspect ratio become larger. The results indicate that a larger direct shear box with a smaller *L*/*H* aspect ratio is more favored for yielding more correct bulk direct shear strength parameters for granular soils because it gives a more global and uniform failure of the entire specimen.

REFERENCES

[1] Itasca, Particle Flow Code (FC2D): User's manual, Itasca, Minneapolis, MN (2002)
[2] Wang, J., Gutierrez, M., Dove, J.E.: Numerical studies of shear banding in interface shear tests using a new strain calculation method. Intl. J. Num. Analy. Meth. Geomech. 31(12), 1349–1366 (2007)
[3] Dove, J.E., Jarrett, J.B.: Behavior of dilative sand interface in a geotribology framework. J. Geotech. Geoenvir. Eng. 128(1), 25–37 (2002)
[4] ASTM D5321-02. Standard test method for determining the coefficient of soil and geosynthetic or geosynthetic and geosynthetic friction by the direct shear method. ASTM, West Conshohocken, PA, USA (1989)
[5] Scarpelli, G., Wood, D.M.: Experimental observations of shear band patterns in direct shear tests. In: Proc. IUTAM Conf. Deformation and Failure of Granular Materials, Delft, pp. 472–484 (1982)
[6] Wang, J., Dove, J.E., Gutierrez, M.: Discrete-continuum analysis of shear band in the direct shear test. Géotechnique 57(6), 513–526 (2007)

VORTICES IN SHEAR BANDS IN SANDS

Sara Abedi* and Amy Rechenmacher

Sonny Astani Department of Civil and Environmental Engineering
University of Southern California
3620 S. Vermont Ave, Los Angeles, CA 90089
* e-mail: abedimas@usc.edu

Summary. We show digital image-based experimental evidence of vortex structures within shear bands in sands deforming in plane strain compression. Spatially varying systematic patterns in kinematic data fields have been found within shear bands, defining a coordinated collapse event of the force chains that initially comprise the shear band upon formation. We have identified co-rotating vortices or "circulation cells" within the shear band by subtracting an overlaid linearly varying shear band displacement field from the observed nonhomogeneous shear band displacement field. These vortex structures correlate precisely with previously observed kinematic patterns.

Keywords: kinematic rotation, nonaffine deformation, shear band, vortex.

1 INTRODUCTION

A distinctive aspect in stressed granular material is the formation of "force chains" or grain columns, the buildup of which resists flow and the collapse of which induces flow [1,3–5]. Nonaffine motion in shear bands is a result of the confined buckling of force chains. Structural development inside the band can be characterized by successive corotating vortices, where buckling force chains can be found through the middle of and around such vortices [7]. According to Williams and Rege [8], vortices, i.e. circulation cells, play a fundamental role in the deformation of granular materials, including the formation of shear bands.

In previous research [6], at the global softening/critical state transition, a distinct spatial pattern among various kinematic quantities was found within the shear band. This pattern was found to be characteristic of a distinct, multi-force chain collapse event, which identified the mesoscale processes responsible for the evolution in bulk shear stress into the critical state. Here, we evaluate the nature of non-affine local displacements within a shear band to investigate the relationship between this force chain collapse event and vortex structures.

2 EXPERIMENTAL METHODS AND ANALYTICAL PROCEDURES

Sand specimens were tested in a plane strain testing apparatus. The prismatic-shaped specimen rests on a low friction sled which helps shear band to grow and propagate with no constraint imposed by boundary conditions [2]. Two rigid glass-lined side walls enforce the zero strain conditions and enable photographic tracking of in-plane shear band deformation [5]. Tests were performed on a mixture of silica sand and a coarse-grained construction sand with D_{50}=0.41 mm. Specimens were encased

R.I. Borja (Ed.): Multiscale and Multiphysics Processes in Geomechanics, SSGG, pp. 21–24.
springerlink.com

in latex membranes, placed in the Plexiglas cell, consolidated anisotropically, and sheared under drained, strain-controlled compression. Throughout shearing, digital images were captured every 0.1% global axial strain.

The digital image correlation (DIC) technique is used to compute the in-plane displacement on the surface of the sand specimen. This technique is based on tracking gray level values within small subsets of pixels between two digital images taken during a deformation process. The software VIC-2D, by Correlated Solutions, Inc., is used to perform the DIC analysis herein.

The DIC method limits the accurate displacement calculation to small strain increments. In order to analyze the kinematics of shear band behavior over large strain increments, the incremental DIC analysis results must be accumulated. Each DIC analysis yields incremental displacements of points on a regular grid referenced to the initial image of that particular increment. Tracking the evolution of a set of material points (Lagrangian frame) over a number of consecutive image increments required interpolation of the displacements. We used cubic spline interpolation method. The deformation gradient, **F**, is then computed from the accumulated displacements. The calculation of area change on the surface of specimen is assumed to be correlated directly with the volume change. The ratio of the volume of an infinitesimal material element to its initial volume is expressed in terms of the jacobian of **F**. Macro-rotations, Ω, that correspond to rigid body rotations are obtained from the polar decomposition of **F**.

3 VORTEX STRUCTURES INSIDE SHEAR BAND

The region above the fully formed shear band translates uniformly downward with the prescribed axial displacement, and the region below the shear band translates laterally with the lubricated base sled. In order to make the lateral displacement below the shear band close to zero, the sled movement during the strain increment of interest is subtracted from the observed displacement field. A rotated coordinate system (x^*, y^*) is utilized, with axis oriented parallel, x^*, and perpendicular, y^*, to the shear band axis. Figure 1a illustrates the nature of the adjusted displacements in the x^*-y^* reference frame over a 0.6% accumulated axial strain increment.

DIC results are in the form of a matrix, consisting of rows and columns. For each column of data the lower and upper boundary of shear band is found and a line of best fit to the data points between these two boundaries is obtained by linear regression. The result is a superimposed affine deformation field inside the shear band. The linear deformation is then subtracted from the observed deformation. A typical result of this analysis is shown in Figure 1(b). Figure 1(b) shows that the upper part of shear band is moving faster than the expected linear field and the lower part is moving slower. In other words, particles are moving in the form shown in Figure 2. However, the rate at which deformation is faster or slower than an affine field is seen to fluctuate nearly periodically along the length of the shear band.

Figure 1(c) shows residual displacement vectors for the displacement increment of Figure 1(a). The lengths of the arrows are proportional to the magnitude of displacement. The existence of well-defined vortices or "circulation cells" is clear in Figure 1(c).

Figure 3(a) through 3(c) shows the relation between the residual displacement field and the fluctuations in obtained kinematic quantities. The centers of vortex structures correlate precisely with the centers of high kinematic rotation (Figure 3(b)). Further, local volumetric contraction is observed at the conflux of adjacent vortices where local displacements from the vortices are in opposition. Slip or "microbands" do not appear to be present at the conflux of adjacent vortices. This is in contrast to [7] and may highlight a key difference between the behavior of real versus idealized granular materials.

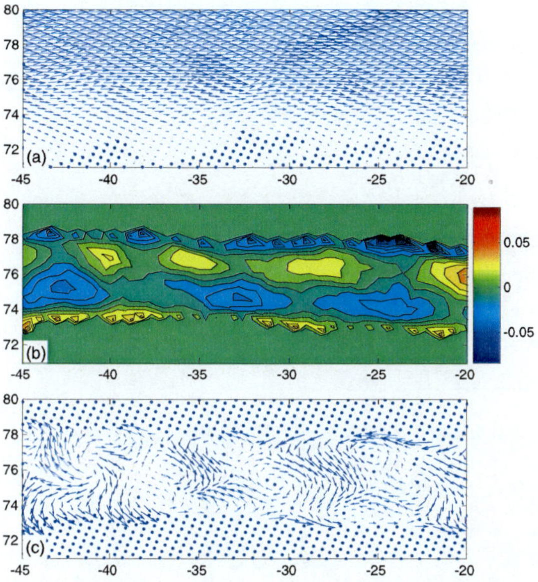

Fig. 1. (a) computed displacement norm after subtracting sled movement from the observed displacement field plotted in the frame of the shear band basis. (b) residual displacement in x* direction after subtracting the linear displacement field from observed displacement field. (c) residual displacement vectors.

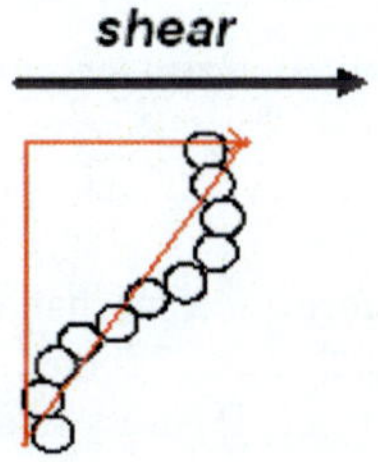

Fig. 2. Form of particle movements in the shear band.

4 CONCLUSIONS

Digital Image Correlation (DIC) is used to measure grain scale displacements in shear bands in sands undergoing plane strain compression. From the resulting fine and continuous displacement field, we have detected co-rotating vortices or "circulation cells" within the shear band. These cells can be hardly seen since they are covered by a superimposed linearly varying shear band displacement field. Once the linear displacement field in the shear band is subtracted from the observed deformation, the cells become clearly visible. The centers of vortex structures are associated with the centers of high kinematic rotation, and local volumetric contraction is observed at the conflux of adjacent vortices where displacements are in opposition. Microbands are not observed between adjacent vortices, highlighting differences in behavior between real and idealized materials.

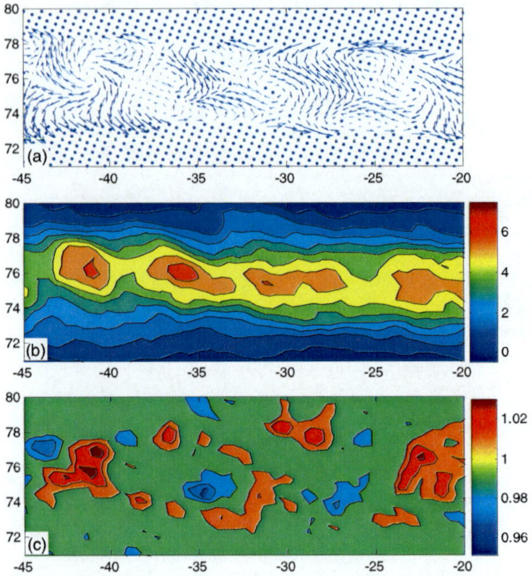

Fig. 3. (a) Residual displacement in x* direction, (b) kinematic rotation, (c) Jacobian (volumetric strain).

REFERENCES

[1] Alonso-Marroquin, F., Vardoulakis, I., Hermann, H.J., Weatherley, D., Mora, P.: Effect of rolling on dissipation in fault gouges. Phys. Rev. E. 74. 31306 (2006)

[2] Drescher, A., Vardoulakis, I., Han, C.: A Biaxial apparatus for testing soils. Geotech. Test. J. 13(3), 226–234 (1990)

[3] Iwashita, K., Oda, M.: Micro-deformation mechanism of shear banding process based on modified distinct element method. Powder. Technol. 109, 192–205 (2000)

[4] Oda, M., Iwashita, K.: Study on couple stress and shear band development in granular media based on numerical simulation analyses. Int. J. Eng. Sci. 38, 1713–1740 (2000)

[5] Rechenmacher, A.L.: Grain-scale processes governing shear band initiation and evolution in sands. J. Mech. Phys. Solids. 54, 22–45 (2006)

[6] Rechenmacher, A.L., Abedi, S., Chupin, O.: Evolution of force chains in shear bands in sands. Géotechnique 60(5), 343–351 (2010)

[7] Tordesillas, A., Muthuswamy, M., Walsh, S.D.C.: Mesoscale measures of nonaffine deformation in dense granular assemblies. J. Eng. Mech. 134(12), 1095–1113 (2008)

[8] Williams, J.R., Rege, N.: Coherent vortex structures in deforming granular materials. Mech. Coh-Frict. Mat. 2(3), 223–236 (1997)

SAND BEHAVIOUR UNDER PRINCIPAL STRESS ROTATION: OBSERVATIONS AND MODELLING

Hai-Sui Yu*, Xia Li, and Yanyan Cai

Nottingham Centre for Geomechanics
Faculty of Engineering, University of Nottingham, NG7 2RD, UK
*e-mail: hai-sui.yu@nottingham.ac.uk

Summary. Loading paths involving large principal stress rotation are frequently encountered in engineering practice. Examples include pavement and railway foundations under traffic loading, offshore foundations, foundations of wind turbine, and earthquake geotechnics. This paper is concerned with sand deformation induced by principal stress rotation. The investigations to be reported include laboratory observations, micro-scale investigation using the discrete element method (DEM), and subsequent development of constitutive models for numerical applications.

Keywords: DEM, fabric tensors, hollow cylindrical apparatus, principal stress rotation, sand plasticity.

1 INTRODUCTION

The effect of principal stress rotation on soil behaviour has attracted much attention in the development of constitutive models. According to classical plasticity theory, the plastic deformation generated during the process of rotational shear would be zero. This theoretical prediction clearly differs from the sand behaviour as observed in the laboratory. A systematic investigation on the deformation of granular material to stress rotation is hence being conducted in the University of Nottingham, and the main findings are summarized here.

2 OBSERVED SAND DEFORMATION DURING STRESS ROTATION

Roscoe [1] was among the first to consider the effect of principal stress rotation on sand behaviour using simple shear tests. It was reported that the principal strain rate direction and the principal stress direction are clearly not identical. This phenomenon is referred to as the non-coaxial soil behavior, which is often observed in situations when the principal stress direction rotates. Many other laboratory investigations on the effect of principal stress rotation were followed using torsional shear or hollow cylindrical apparatus [2-4]. All experimental test results confirmed the observations of non-coaxial behaviour and showed that the degree of non-coaxiality reduces with increasing stress ratios. Significant volume contraction has also been observed as another consequence of principal stress rotation. Most recently, Cai [5] conducted pure stress rotation tests on Portaway and Leighton Buzzard (Fraction B) sands of different densities. The results show that the degree of non-coaxiality is more pronounced in dense specimen than that in medium dense specimen, also stronger evidence for noncoaxiality can be found in

R.I. Borja (Ed.): Multiscale and Multiphysics Processes in Geomechanics, SSGG, pp. 25–28.
springerlink.com

Leighton Buzzard sand than in Portaway sand, while the difference diminishes as the shear stress level increases. The degrees of non-coaxiality are higher, about 38°, at the beginning of shearing, and decrease within a range of 10°–20° as shear continues.

3 PARTICLE-SCALE INVESTIGATION USING THE DISCRETE ELEMENT METHOD

DEM provides valuable information not only on macro-scale about stress and strain behaviour, but also on micro-scale of contact forces and particle displacements. A series of numerical experiments of pure stress rotations have been conducted and the results are presented [6]. The numerical results confirmed laboratory observations on the non-coaxial and contractive behaviour as reported in the literature. Based on the knowledge of internal soil structure, the volume contraction has been explained as a result of the decrease in the internal structure size. The non-coaxiality is found to be induced by the fabric anisotropy related to internal structure rotation during the principal stress rotation.

4 DEVELOPMENT OF CONSTITUTIVE MODELS

Soil deformation induced by stress rotation is closely associated with material anisotropy. Research has been carried out to develop a general microstructure-based plasticity theory in terms of both stress and fabric tensors.

4.1 Microstructure-Based Plasticity Theory

Yu [7], for example, proposed a general approach by assuming that at yielding, the plastic strain rate tensor consists of two components. The first part is determined by a plastic potential as in the conventional plasticity theory. The second part is assumed to be related to the rate of the fabric tensor. The proposed plastic flow rule is of the general form:

$$\dot{\epsilon}^p_{ij} = \dot{\lambda}\frac{\partial g}{\partial \sigma_{ij}} + A(F_{ij})\dot{F}_{ij}, \tag{1}$$

where $A(F_{ij})$ is a material function, which may depend on the stress tensor, but it may also be assumed to be a constant for some cases. The proposed plastic flow rule is consistent with the experimental findings under continuous principal strain rotation where it was observed that the fabric rate tensor appears to be proportional to the strain rate tensor.

Both the yield function and plastic potential are assumed to be functions of the fabric tensor. The yield function is assumed to depend on the stress tensor, the fabric tensor, and a hardening parameter:

$$f(\sigma_{ij}, F_{ij}, k) = 0, \tag{2}$$

where the hardening parameter k is generally assumed to be a function of plastic strains. The yield function will generally be an anisotropic function in the stress space. The corresponding general plastic potential may read

$$g(\sigma_{ij}, F_{ij}, k) = 0. \tag{3}$$

Micromechanical considerations suggest a strong relationship between the fabric tensor and the deviatoric stress or shear stress ratio. We therefore assume that the fabric tensor evolution law is of the following general form:

$$\dot{F}_{ij} = \eta B(s_{ij}, \dot{s}_{ij}), \tag{4}$$

where $s_{ij} = \sigma_{ij} - \sigma_{kk}\delta_{ij}/3$ is the deviatoric stress tensor, B is a material function, and η is a material constant.

4.2 Relations with Existing Plasticity Models

With certain fabric evolution laws, the proposed flow rule Eq.(1) reduces to some well-known existing non-coaxial models. In other words, the present fabric-based plastic flow theory provides a new micromechanical interpretation of these non-coaxial models.

The present plastic flow rule Eq.(1) reduces to that of the yield vertex model proposed by [8] with a Drucker-Prager yield criterion if we assume

$$A(F_{ij}) = \frac{1}{\eta h_1}, \qquad \dot{F}_{ij} = \eta\left(\dot{s}_{ij} - \frac{s_{ij}s_{kl}}{2\tau^2}\dot{s}_{kl}\right), \qquad \tau = \sqrt{s_{ij}s_{ij}/2}, \tag{5}$$

where h_1 is a plastic modulus.

The non-coaxial plastic flow rules of [9-10] may be obtained by using the following fabric evolution law:

$$\dot{F}_{ij} = \eta\left[\dot{s}_{ij} - \left(\dot{s}_{kl}n_{kl}\right)n_{ij}\right], \qquad \text{where} \qquad n_{ij} = s_{ij}/(\sqrt{2}\tau) \tag{6}$$

In the theory of [11], the fabric tensor is assumed to vary with a deviatoric stress ratio

$$\dot{F}_{ij} = \eta(\dot{s}_{ij}/\sigma_{kk}) = \eta\left[\frac{\dot{s}_{ij}}{\sigma_{kk}} - \frac{s_{ij}}{(\sigma_{kk})^2}\dot{\sigma}_{kk}\right]. \tag{7}$$

In the micromechanics models, it was assumed that the fabric tensor is proportional to the stress tensor normalised by the mean pressure. Therefore, we have

$$\dot{F}_{ij} = \eta(\dot{\sigma}_{ij}/\sigma_{kk}) = \eta\left[\frac{\dot{\sigma}_{ij}}{\sigma_{kk}} - \frac{\sigma_{ij}}{(\sigma_{kk})^2}\dot{\sigma}_{kk}\right]. \tag{8}$$

The well-known double shearing theory, initially proposed by [12] and later generalised by [13-14], can be obtained by assuming that the fabric tensor changes with the change of the direction of principal stresses, namely,

$$\dot{F}_{ij} = \frac{\eta}{\tau}\left(\dot{s}_{ij} - \frac{s_{ij}s_{kl}}{2\tau^2}\dot{s}_{kl}\right) \tag{9}$$

5 SUMMARY

Using laboratory tests and DEM numerical experiments, this paper is concerned with modelling sand deformation characteristics under principal stress rotation. The volume contraction and non-coaxiality are two key features that have been observed in the laboratory and confirmed by DEM simulations. A new micromechanical fabric-based plasticity theory has been proposed as a general theoretical framework for developing realistic non-coaxial stress-strain behaviour of granular materials. Several well-known existing micromechanics and non-coaxial plasticity models are shown to be special cases of the proposed fabric-based plasticity theory.

REFERENCES

[1] Roscoe, K.H.: The influence of strains in soil mechanics. Géotechnique 20, 129–170 (1970)

[2] Hight, D.W., Gens, A., Symes, M.J.: The development of a new hollow cylinder apparatus for investigating the effects of principal stress rotation in soils. Géotechnique 33, 355–383 (1983)

[3] Gutierrez, M., Ishihara, K., Towhata, I.: Flow theory for sand rotation of principal stress direction. Soils and Foundations 31, 121–132 (1991)

[4] Yang, Z.X., Li, X.S., Yang, J.: Undrained anisotropy and rotational shear in granular soil. Géotechnique 57, 371–384 (2007)

[5] Cai, Y.: An experimental study of non-coaxial soil behaviour using hollow cylinder testing. Ph.D Dissertation in Department of Civil Engineering, University of Nottingham, Nottingham (2010)

[6] Li, X., Yu, H.S.: Numerical investigation of granular material behavior under rotational shear. Géotechnique 60, 381–394 (2010)

[7] Yu, H.S.: Non-coaxial theories of plasticity for granular materials. In: The 12th International Conference of International Association for Computer Methods and Advances in Geomechanics (IACMAG), India, Goa (2008)

[8] Rudnicki, J.W., Rice, J.R.: Condintions for the localisation of deformation in pressure-sensitive dilatant materials. Journal of the Mechanics and Physics of Solids 23, 371–394 (1975)

[9] Tsutsumi, S., Hashiguchi, K.: General non-proportional loading behaviour of soils. International Journal of Plasticity 21, 1941–1969 (2005)

[10] Yang, Y., Yu, H.S.: A non-coaxial critical state soil model and its application to simple shear simulations. International Journal for Numerical and Analytical Methods in Geomechanics 30, 1369–1390 (2006)

[11] Wan, R.G., Guo, P.J.: Stress dilatancy and fabric dependencies on sand behaviour. Journal of Engineering Mechanics, ASCE 130, 635–645 (2004)

[12] Spencer, A.J.M.: Deformation of ideal granular materials. In: Sewell, H.A. (ed.) Mechanics of Solids, pp. 607–652 (1982)

[13] Harris, D.: Constitutive equations for planar deformation of rigid plastic materials. Journal of the Mechanics and Physics of Solids 41, 1515–1531 (1993)

[14] Yu, H.S., Yuan, X.: On a class of non-coaxial plasticity models for granular soils. Proceedings of the Royal Society London, Series A 462, 725–748 (2006)

PARAELASTIC DEFORMATIONS IN HYPOPLASTICITY

Andrzej Niemunis[1,*] and Felipe Prada[1,2]

[1] Institute of Soil Mechanics and Rock Mechanics
Karlsruhe Institute of Technology
Engler-Bunte Ring 14, Gebäude 40.22, 76131 Karlsruhe, Germany
[*]e-mail: andrzej.niemunis@ibf.uni-karlsruhe.de,
web page: http://www.rz.uni-karlsruhe.de/~gn99/
[2] Department of Civil Engineering
Universidad de Los Andes
Bogotá, Colombia
e-mail: felipe.prada@ibf.uni-karlsruhe.de

Summary. Paraelasticity describes a family of multiaxial hysteretic models. They preserve stress but dissipate energy upon closed strain cycles. Paraelasticity in combination with hypoplasticity is discussed.

Keywords: constitutive model, hypoplasticity, hysteresis, paraelasticity.

1 INTRODUCTION

In constitutive modeling of soils it is convenient, at least conceptually, to distinguish between small, intermediate and large strain amplitudes. A constitutive model can be designed to interpolate between the conservative (e.g. hyperelastic), hysteretic (e.g. paraelastic) and cumulative (e.g. hypoplastic) behaviour, respectively.

The currently used hypoplastic models do not make such distinction. Their major disadvantage is the excessive ratcheting during asymmetric cycles. Parallel coupling of several hypoplastic elements [5] does not satisfactorily mitigate the problem. Moreover, contrarily to experimental evidence, small closed oval strain cycles generate "hypoelastic" relaxation, whose direction depends on the sense of circulation. Alternatively multi-surface plasticity [3, 4, 1] is applicable for small cycles. However, it does not guarantee full reversibility. Other authors make use of the equivalence between hysteretic and viscous damping in the range of intermediate amplitudes. This equivalence holds for a particular frequency only. The so-called intergranular strain (IS-)hypoplasticity was meant to improve the original hypoplasticity for small-strains. Unfortunately, this new IS-hypoplasticity could neither provide full reversibility nor obey the Masing rules.

The paraelasticity was first proposed by Hueckel and Nova (H-N) in 1979 [2] as a 3d generalization of 1d hysteretic models. Here, a combination of hypoplasticity with paraelasticity is proposed to supersede the IS-hypoplasticity. Moreover, we introduce a remedy against an important defect of the H-N paraelasticity. The incrementally linear part of hypoplasticity is made paraelastic and the linear part thereof is hyperelastic.

R.I. Borja (Ed.): Multiscale and Multiphysics Processes in Geomechanics, SSGG, pp. 29–32.
springerlink.com

2 FROM 1D TO 3D HYSTERETIC MODEL

The hysteretic relation between stress and strain is not unique. Unique stresses are obtained at the reversals of the strain path only. Despite various mathematical formulations all 1d hysteretic models are quite similar: they introduce reversals, distinguish between loading and unloading, satisfy the Masing rules and reproduce the well known characteristic curves describing stiffness $G(\gamma^{\text{ampl}})$ and damping ratio and $D(\gamma^{\text{ampl}})$. Unfortunately, their extension to the full 3d case is not straightforward:

1. The linear portion is pressure dependent and requires an elastic potential.
2. Definition of loading and unloading requires a definition of loading direction.
3. A problem of discontinuity between loading and unloading arises in a similar manner as in elastoplasticity.
4. A system of bookkeeping of the reversal points becomes complex.
5. Upon an arbitrary closed strain path a complex system of sub-reversals may be generated and subsequently deleted without violating the stress reversibility.

The idea of the current formulation is to distinguish between paraelastic and cumulative effects. The latter are delegated to the hierarchically superposed hypoplastic model.

3 IMPROVEMENTS TO THE H-N PROPOSAL

The essential concept of the H-N paraelasticity is the nonlinear secant compliance (or stiffness). It is preserved in the current formulation. However, upon some strain paths, the H-N model causes jumps in stress and therefore an improvement was necessary.

3.1 Reversals

A reversal appears upon a strain path when the distance of the current strain to the strain of the most recent reversal starts to decrease. We shall demonstrate, however, that the H-N distance $\bar{d}_{AR} = \|\boldsymbol{\epsilon} - \boldsymbol{\epsilon}_R\|$ is unsuitable, see Fig. 1. The paraelastic secant stiffness interrelates stress and strain portions measured from the most recent reversal defined by the triple $R_1 = \{\boldsymbol{\epsilon}^{R1}, \boldsymbol{\sigma}^{R1}, \mathbf{N}^{R1}\}$ describing the reversal strain, stress and loading direction. Earlier reversals are $R_2 = \{\boldsymbol{\epsilon}^{R2}, \boldsymbol{\sigma}^{R2}, \mathbf{N}^{R2}\}$, etc. The current state is described by $A = \{\boldsymbol{\epsilon}, \boldsymbol{\sigma}, \mathbf{N}\}$. The loading "circle" (a 6d hypersphere) passes through $\boldsymbol{\epsilon}^{R1}$ and $\boldsymbol{\epsilon}$ and has the unit outer normal $\mathbf{N}^{R1}$ at $\boldsymbol{\epsilon}^{R1}$, Fig. 1.

3.2 Distance

The novel definition of distance from A to R_1 is the diameter of the loading circle, Fig. 1,

$$d_{AR_1} = 2r = -\frac{\mathbf{e} : \mathbf{e}}{\mathbf{e} : \mathbf{N}^{R1}} \geq 0 \quad \text{with} \quad \mathbf{s} = \boldsymbol{\sigma} - \boldsymbol{\sigma}^{R1} \quad \text{and} \quad \mathbf{e} = \boldsymbol{\epsilon} - \boldsymbol{\epsilon}^{R1}. \tag{1}$$

Analogously, the distance between R_1 and R_2 can be estimated. Distances between non-consecutive reversals, e.g. between A and R_2, need not be defined.

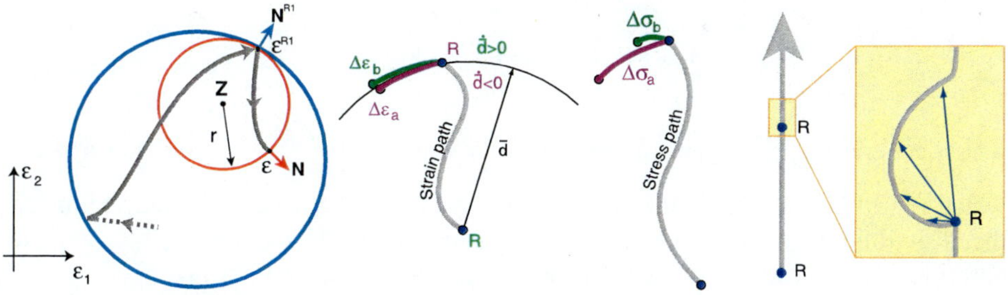

Fig. 1. Left: Loading circle (small) and reversal circle (large) in the strain space generated by the strain path with reversals. Middle and Right: Shortcomings of the H-N definition $\bar{d}_{AR} = \|\mathbf{e}\|$ mitigated by (1)

If d_{AR_1} increases monotonically ($\dot{d}_{AR_1} > 0$, loading) the list of reversals does not change. The *push*-operation is undertaken if d_{AR_1} starts to decrease ($\dot{d}_{AR_1} < 0$, unloading). The current state A (before increment) and the reversal R_1 generate a *reversal circle*. We renominate $R_1 := A$ and the current increment develops a new small loading circle from the new R_1, Fig. 2, cases 1,2 and 3UL. It may also happen that the current loading circle outgrows the most recent reversal circle $d_{AR_1} > d_{RR_1}$). In this case the most recent reversal R_1 is removed (=*popped*) as depicted in Fig. 2, case 3L.

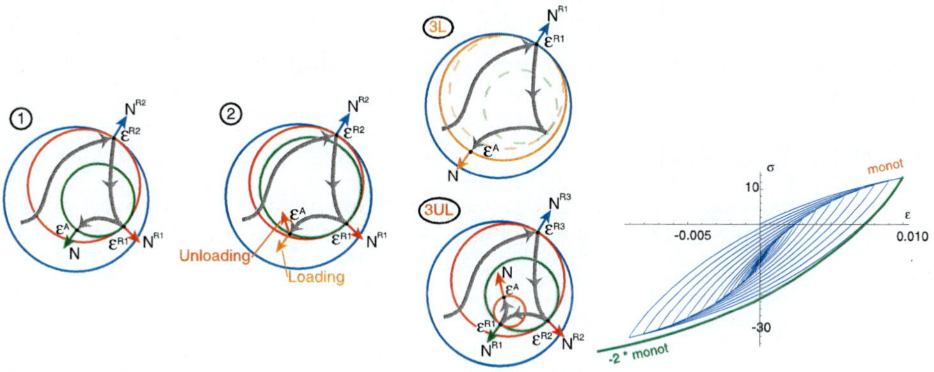

Fig. 2. Left: unloading (1,2,3UL) creates new reversals. (Over)loading (3L) deletes a reversal. Right: for the special case of 1d strain cycles Masing rule is obeyed.

3.3 Constitutive Equations

The nonlinear secant stress–strain relationship is

$$\mathbf{s} = \mathsf{H} : \mathbf{e} = (1 - f d_{AR_1}^{\chi})\mathsf{E} : \mathbf{e}, \tag{2}$$

wherein f and χ are material constants. In the simplest case we may take $\mathsf{E} = \lambda\, \mathbf{1}\, \mathbf{1} + 2\mu \mathsf{I}$ with the Lamé constants λ and μ. For geomaterials a barotropic increase of stiffness is observed. It requires a formulation of an elastic potential, for example

$$\bar{\psi}(\mathbf{T}) = c\,(-\mathrm{tr}\,\mathbf{T})^{\alpha}\,\|\mathbf{T}\|^{2-n-\alpha} \tag{3}$$

which generates a homogeneous stiffness $\mathsf{E}(\mathbf{T}) = \left[\dfrac{\partial^2\bar{\psi}(\mathbf{T})}{\partial\mathbf{T}\partial\mathbf{T}}\right]^{-1}$ of order n and for which $\boldsymbol{\epsilon}^{\mathrm{el}}(\mathbf{T})$ is analytically invertible. The parameters n, c, α can be chosen to match the desired homogeneity, Poisson ratio, and stress-induced anisotropy. The complete description of the model will be published in Acta Geotechnica in 2011.

4 COUPLING OF PARAELASTICITY AND HYPOPLASTICITY

Our paraelastic model introduces the outermost reversal circle which cannot be surpassed. Its diameter corresponds to the maximal double amplitude d_L for which paraelasticity works. Upon larger cycles we apply *dragging*: all but one reversal are deleted and the circle is moved along with the current state similarly as in kinematic hardening. The stress rate is calculated from the hypoplastic model.

Formally, paraelasticity and hypoplasticity are parallel coupled and their stress increments are added. Given $\Delta\boldsymbol{\epsilon}$ their contributions are weighted depending on the diameter d_{AR_1}

$$\begin{aligned}\Delta\boldsymbol{\sigma} &= (\mathsf{L} - H_L(1-w)) : \Delta\boldsymbol{\epsilon} + N\|\Delta\boldsymbol{\epsilon}\|w + \Delta\boldsymbol{\sigma}^{PE} \\ &\text{with} \quad w = \left\langle \frac{d_{AR_1} - d_0}{d_L - d_0} \right\rangle^{\beta_R} \quad \text{and} \quad H_L = H(d_L).\end{aligned} \tag{4}$$

The purpose of this approach is to provide a smooth transition between paraelasticity and hypoplasticity.

ACKNOWLEDGMENTS

The financial support from the German Research Community (DFG-GeoTech 1136) and from COLCIENCIAS, Universidad de Los Andes/CEIBA is gratefully acknowledged.

REFERENCES

[1] Dafalias, Y.F.: Bounding surface elastoplasticity-viscoplasticity for particulate cohesive media. In: Proceedings IUTAM Conf. Delft.Deformation and Failure of Granular Materials, pp. 97–107 (1982)
[2] Hueckel, T., Nova, R.: Some hysteresis effects of the behaviour of geologic media. International Journal of Solids and Structures 15, 625–642 (1979)
[3] Iwan, W.D.: A distributed-element model for hysteresis and its steady-state dynamic response. Journal of Applied Mechanics 33, 893–900 (1966)
[4] Mróz, Z., Norris, V.A., Zienkiewicz, O.C.: An anisotropic, critical state model for soils subject to cyclic loading. Géotechnique 31(4), 451–469 (1981)
[5] Valanis, K.C.: A theory of viscoplasticity without a yield surface, Part I. General theory. Archiwum Mechaniki Stosowanej 23, 517–533 (1971)

WHEN FAILURE IS AN OPTION: REWIRING GRANULAR NETWORKS

Antoinette Tordesillas

Department of Mathematics and Statistics
University of Melbourne
Victoria 3010 Australia
e-mail: atordesi@ms.unimelb.edu.au, web page: http://www.mgm.ms.unimelb.edu.au

Summary. We report on recent advances in the multiscale analysis of emergent networks in deforming, dense granular assemblies. Conceptually, we envision the material as an evolving mathematical graph or network that is employing specific rules in adapting its contact and contact force networks in response to applied loads. In this context, the question that arises is: Is the granular material re-wiring its networks to optimize a particular measure and, if so, what is this measure? We probe this question within a multidisciplinary framework. Synergistic blends of techniques from Structural Mechanics, Complex Networks and Dynamical Systems are employed to uncover clues on self-organized pattern formation in the lead up to and during failure. We demonstrate these techniques by analysing data from various compression tests, both from discrete element simulations and physical experiments on photoelastic disks.

Keywords: cluster conformations, granular materials, motifs, networks, stability.

1 INTRODUCTION

One of the great challenges in the science of *complex materials* — materials capable of emergent or self-organized pattern formation — is deciphering their inherent structural design principles as they deform in response to external loads. Herein, we introduce new mathematical tools for the multiscale analysis of geomaterials. These tools offer unique capabilities for comprehensive characterization and exploration of: topology, stability, dynamics and function; their interconnections; and their evolution during the material's loading history. Particle scale information from experimental measurements and discrete element simulations is key to this effort [1–3]. Results from studies of idealized granular materials suggest that not only are these tools effective individually at uncovering the governing physics of material behavior, but also more potent when overlaid and/or fused together for certain problems [4–7, 9].

In this presentation, we focus on tools that spawn from a representation of the granular material as a mathematical graph or network comprising nodes (or vertices) and links (or edges). Networks underpin virtually all complex systems. Arguably one of the best examples where networks feature prominently is in deforming dense granular systems — with connectivity and force transmission being governed by the contact and contact force network, respectively. Specifically, we may cast the particles as the nodes of a graph and links between two nodes exist where there is a physical contact between the corresponding particles [4]. Labeling the nodes by an index $i = 1, ..., N$,

R.I. Borja (Ed.): Multiscale and Multiphysics Processes in Geomechanics, SSGG, pp. 33–36.
springerlink.com

the network may be succinctly summarized by the so-called adjacency matrix $\mathbf{A}$, which has entries $A_{ij} = 1$ if there is a link from node i to node j and $A_{ij} = 0$ otherwise. The quasi-statically deforming material may then be analyzed as an evolving graph or a sequence of graphs each of which corresponds to a particular state in the loading history of the material. Many types of smaller sub-graphs can be extracted from the global network and their evolution tracked throughout loading. The unweighted graph, as defined by $\mathbf{A}$, carries information on material connectivity alone. However, we might consider an assortment of weighted graphs in which the links carry another piece of information (weight w_{ij}): e.g. magnitude of the contact force. In this case, the adjacency matrix $A_{ij} = w_{ij}$ if a link exist between nodes i and j, and zero otherwise. A spectral analysis of this matrix and its associated normalized Laplacian matrix may then be undertaken to extract various metrics that characterize the graph and its evolution.

2 QUANTIFYING CONNECTIVITY VIA COMPLEX NETWORKS

Complex Networks is a relatively new direction in granular media research. This is surprising given the proven relevance of connectivity (i.e. fabric) and associated force networks (e.g. force chains) in governing the bulk rheology of these materials [4]. The evolving loss of connectivity in the material domain of many forms of complex materials has been long touted as key to understanding dissipation and plastic deformation. Loss of inter-particle contacts, slip, dislocations and other types of damage processes that cause breakdown in the connectivity are believed to be the origin of material irreversibility. These processes involve a loss of free energy and are thus a primary source of dissipation. More importantly, a fundamental tenet of Continuum Theory is that materials undergo an affine deformation and yet the associated deformation of these important processes is often highly non-affine.

We have explored the efficacy of methods from Complex Networks as a means to identify suitable metrics that quantify the evolution of connectivity and structure at multiple length scales [4]. Key events that cause loss of connectivity within the material domain, e.g. force chain buckling [10], have been investigated. We discuss how network centrality measures, its k-core architecture, etc. relates to such dissipative and non-affine processes as well as to corresponding trends observed in the evolving macroscopic network [4, 6, 7]. Applying these techniques to "growing networks" in systems where particle breakage occurs along with force chain buckling (e.g. [8]) is also proving to be fertile ground for studies of scale invariance of network properties.

3 STRUCTURAL STABILITY OF GRANULAR MOTIFS

Motifs in networks or graphs are connected *sub*-graphs that occur more frequently than in random networks. In a deforming granular material, these motifs represent prevalent and persistent mesoscopic particle clusters that may be viewed as the basic building blocks of granular matter [5, 9]. Examples of these include (i) force chains and (ii) particles forming minimal cycles (e.g. three particles in mutual contact forming 3 cycles or triangles). The former have received significant attention owing to their fundamental role in force transmission, while the relevance of the latter has been recently demonstrated in [5].

The existence of motifs suggests a strong preference of the material to form particular mesoscopic structures or conformations throughout loading [5]. The question that then arises is: why are such structures preferred over a multitude of other accessible configurations? Even more tantalizing is nascent evidence which shows these preferred structures co-evolve in ways suggestive of organized and cooperative behavior. Do these hold clues to functional activity? An understanding of this co-evolution may hold the secrets to a granular material's capacity to resist and maintain stability under load. Viewed from this perspective, the most obvious property to examine is structural stability [9]. Structural stability is not only related to material properties but also to topology and boundary conditions. Here, a structural mechanics approach is employed, first, to devise a new stability measure for such conformations in equilibrium and, second, to use this measure to explore the evolving stability of specific cluster conformations, i.e. particles forming force chains and minimal cycles.

When network properties are overlaid with stability measures, clues to functional activity are uncovered. In a biaxial test, mesoscopic structures reminiscent of "power towers" form to resist the compressive load. The strong columns of the tower are the force chains, while the trusses are the strong 3 cycles (termed 3-force cycles) [5]. Past studies have shown that the failure of the granular material is presaged by buckling force chains, eventually leading to the development of macroscopic shear bands [1, 10]. The force chains typically fail by buckling through large relative rotations between constituent particles. The role of the 3-force cycles is to secure and reinforce the force chain columns by frustrating rotations as well as providing strong lateral support. During the period of shear band development, i.e. from the onset of to the fully developed shear band (from just prior to peak shear stress to the start of the critical state regime), the granular material calls forth these reinforcing 3-force cycles, creating power towers throughout the shear band region. However, with the combined effect of increasing compression and dilatation, the material's attempts to maintain stability ultimately proves futile.

4 DYNAMICAL SYSTEMS APPROACH TO GRANULAR CONFORMATIONAL TRANSITIONS

Particle cluster conformations and their patterns of transition from one conformation to another can be studied, along with their structural stabilities, using a Dynamical Systems approach [11]. Starting from the graph or network representation of the material, we model the evolution of particle clusters in the deforming material using Markov transition matrices. We identify sets of conformations which are almost-invariant under the dynamics, quantify transition probabilities and map out the global dynamics within the framework of a Perron-Frobenius (P-F) formalism. We find that the material exhibits preferred cluster conformations and, incredibly, these configurations can be directly related to the granular motifs found in the Complex Networks analysis. Along similar lines to molecular energy landscapes, we have also explored these conformational transition dynamics from the standpoint of structural stability. We find evidence for preferred stability states and, remarkably, transition patterns between stability states follow those of the granular motifs. That is, cluster conformations with high clustering and stability values are more likely to transition into other conformations with similarly high clustering and stability values. Likewise, conformations with low clustering and stability values are more likely to transition into conformations with low clustering and stability values.

5 CONCLUSIONS

Contemporary multidisciplinary research, embodying techniques of Complex Networks, Dynamical Systems and Structural Mechanics, herald new directions in the multiscale analysis of geomaterials. Preliminary studies show promise and, for certain problems, the fusion of these techniques can potentially crystallize understanding of granular materials at unprecedented levels of detail.

REFERENCES

[1] Tordesillas, A.: Force chain buckling, unjamming transitions and shear banding in dense granular assemblies. Philosophical Magazine 87, 4987–5016 (2007)

[2] Tordesillas, A., Zhang, J., Behringer, R.P.: Buckling force chains in dense granular assemblies: physical and numerical experiments. Geomechanics and Geoengineering 1, 3–16 (2010)

[3] Zhang, J., Majmudar, T., Tordesillas, A., Behringer, R.P.: Statistical properties of a 2D granular materials subjected to cyclic shear. Granular Matter, 12 (2010), doi:10.1007/s10035-010-0170-2

[4] Walker, D.M., Tordesillas, A.: Topological evolution in dense granular materials: A complex networks perspective. Int. J. Solids Struct. 47, 624–639 (2010)

[5] Tordesillas, A., Walker, D.M., Lin, Q.: Force cycles and force chains. Phys. Rev. E 81, 011302 (2010)

[6] Walker, D.M., Tordesillas, A., Thornton, C., Behringer, R.P., Zhang, J., Peters, J.F.: Percolating contact subnetworks on the edge of isostaticity (in review)

[7] Tordesillas, A., O'Sullivan, P., Walker, D.M.: Paramitha and evolution of functional connectivity in contact and force chain networks: feature vectors, k-cores and minimal cycles. In: CRAS Proceedings of the French Academy of Science (in press)

[8] Ben-Nun, O., Einav, I., Tordesillas, A.: Force attractor in confined comminution of granular materials. Phys. Rev. Lett. 104 (2010)

[9] Tordesillas, A., Lin, Q., Zhang, J., Behringer, R.P., Shi, J.: Structural stability and jamming of self-organized cluster conformations in dense granular materials. J. Mech. Phys. Solids (to appear)

[10] Tordesillas, A., Muthuswamy, M.: On the modelling of confined buckling of force chains. J. Mech. Phys. Solids 57, 706–727 (2010)

[11] Tordesillas, A., Walker, D., Froyland, G., Zhang, J., Behringer, R.P.: Stability landscapes and transition dynamics of granular motifs (in preparation)

BIFURCATIONS FROM THE MICROSCALE TO THE MACROSCALE: A NEW PARADIGM FOR MODELING AND EXPERIMENTS

René Chambon

Laboratoire 3S-R
University of Grenoble, Université Joseph Fourier
BP 53 38041 Grenoble France
e-mail: Rene.Chambon@hmg.inpg.fr

Summary. Loss of uniqueness of the solutions of initial boundary value problems obtained by the use of constitutive equations incorporating strength degradations can be observed numerically even with enhanced models. As engineers, we must address this issue. In this paper the problem is studied both at the macro level and the micro level, in conjunction with some experimental data and theoretical results. It is concluded that the situation has to be studied differently at the element and structural levels. In both cases, however, this has to prompt us to change some of our routine ideas, which could mean abandoning the classical paradigm of continuum mechanics.

Keywords: bifurcation, degradation, enhanced models, experimental data, numerical modeling.

1 INTRODUCTION

The most commonly used method for solving new geomechanical problems is based on continuum mechanics. Mathematically this leads to an initial boundary value problem, since it is necessary to follow the loading paths induced by the history of boundary conditions. The constitutive equations (more precisely, the parameters of the constitutive equations) are obtained by experiments assumed to be homogeneous and measurements at the boundaries of these experiments. Unfortunately, geomaterials often exhibit a clear degradation of strength which renders an accurate modeling close to rupture difficult. Especially we have good reason to think that in this case, the solutions of the mathematical problem may not be unique. A new and promising way to obtain some knowledge about constitutive equations is provided by the so-called FE^2 method. In this case an element sample is modeled as a structure and the response of this element is used as a numerical constitutive equation. At the element level the same kind of difficulties as for structures may be encountered, if some degradation of strength is included in the model. In order to clarify this discussion, it is necessary to distinguish between the mathematical problem and the numerical one.

2 GENERAL FORMULATION OF THE PROBLEMS

For simplicity, in this section equations are written for a classical monophasic problem. The use of enhanced models (such as the local second gradient models) and coupling does not change the form of the following.

R.I. Borja (Ed.): Multiscale and Multiphysics Processes in Geomechanics, SSGG, pp. 37–40.
springerlink.com

2.1 Initial Boundary Value Problem

Let a body which at time t is in Ω^t, the boundary of which is denoted Γ^t. Balance equations imply that for any kinematically admissible field $\dot{u}^*$

$$\int_{\Omega^t} \sigma^t_{ij} \frac{\partial \dot{u}^*_i}{\partial x^t_j} \, d\Omega^t - \int_{\Gamma^t_\sigma} t^t_i \dot{u}^*_i \, d\Gamma^t - \int_{\Omega^t} \rho^t f^t_i \dot{u}^*_i \, d\Omega^t = 0 \tag{1}$$

where Γ^t_σ is the part of Γ^t where the stress is prescribed. The other part where the displacments are prescribed is denoted Γ^t_u, σ^t_{ij} are the components of the Cauchy stress tensor, ρ^t is the mass density per unit volume and f^t_i are the body forces per unit mass.

Moreover, the constitutive equation has to be defined as a function of the history of the whole kinematics:

$$\sigma^t_{ij} = \mathcal{F}_{ij}[\frac{\partial x^\tau_m}{\partial x^0_p}(\tau = 0 \rightarrow t)] \tag{2}$$

The initial boundary value problem can be defined as follows. The history of boundary conditions (stresses t^t_i on Γ^t_σ and displacements on Γ^t_u) as well as body forces are prescribed. The initial state is known and satisfies the equilibrium condition, and we are looking for the history of the positions of the body Ω^t and of the stress field σ^t_{ij}.

2.2 Rate Problem

The rate problem can be definned as follows: We know the position Ω^t of a body, we are looking for the velocity field defined here by $\dot{\varepsilon}_{jl}$ (the strain rate) and $\dot{\omega}_{li}$ (the spin), such that for any kinematically admissible field $\dot{u}^*$:

$$\int_{\Omega^t} [\tilde{\sigma}_{ik} - \sigma^t_{lk}\dot{\omega}_{li} + \sigma^t_{ik}\dot{\varepsilon}_{ll} - \sigma^t_{ij}\dot{\varepsilon}_{kj}] \frac{\partial \dot{u}^*_i}{\partial x^t_k} \, d\Omega^t - \int_{\Gamma^t_\sigma} \dot{t}^t_i \dot{u}^*_i \, d\Gamma^t - \int_{\Omega^t} \rho^t \dot{f}^t_i \dot{u}^*_i \, d\Omega^t = 0, \tag{3}$$

where $\dot{t}^t_i$ is prescribed on Γ^t_σ and the velocity is prescribed on Γ^t_u; $\tilde{\sigma}_{ij}$ is the Jaumann derivative of the Cauchy stress. The rate type constitutive equation is usually given by

$$\tilde{\sigma}_{ij} = \mathcal{G}_{ij}(\frac{\partial \dot{u}_k}{\partial x^t_j}), \tag{4}$$

where $\mathcal{G}_{ij}$ is a non-linear function.

3 MATHEMATICAL RESULTS

To our knowledge, there is no general mathematical result available for geomaterials (i.e. for constitutive equation involving, broadly speaking, non-associativeness) for the initial boundary value problem.

However, some results are now known for the rate problem. A general condition that provides existence and uniqueness of the solution of this problem is related to some monotony properties (see [1] and [2] for more details and for the proofs). But it is interesting to recall here a very important result. For classical elastoplasticity (which means one mechanism flow theory of plasticity) and for hypoplastic model (which means models obeying the formulae recalled in [2]), if a state is such that everywhere the second-order work is positive, then the solution of the rate problem exists and is unique. If this condition is not met, then there is no proved result.

Among the class of rate boundary value problems some are more simple, the ones for which the state is homogeneous and for which the (rate) boundary conditions allow at least an homogeneous solution. Two classical studies belong to this class: Rice [3] analysis (which has in general nothing to do with the so-called criterion about the acoustic tensor) for which is searched another solution involving a (not necessarily shear) band and Nova [4] controllability for which alternative solutions are chosen to be homogeneous. According to this way of thinking some general (independent of the constitutive equation) results can be proved [5]. Non-positiveness of the second-order work implies non-uniqueness. Due to non-associativeness of the constitutive equations used for geomaterial this is encountered before the classical failure condition.

4 NUMERICAL RESULTS

Without theoretical results, it is, however, possible to have some insight into this very important problem. Many examples of such situations are available. See, for instance, some results for the orientation of localized bands, in [6] and [7].

The loss of internal length in classical model is often assumed to be the origin of such difficulties. Unfortunately, this is not the case as it has been exemplified for the so-called local second-gradient models. See [8] and [9] for biaxial tests. But such results are not only observed for problem starting for initial homogenous states, they can also be observed for more complex problem, the borehole problem for instance (see [10]).

When modeling geomaterials as multi-scale structures using the so-called Finite Element square method, for instance, if some degradation process is assumed at the micro-scale, loss of uniqueness is also observed at this scale (see [11], for instance).

5 CONCLUSIONS

This kind of results cannot be ignored, even if doing routinely computations (especially with computer codes which give always a solution), cannot warn us, so we have a priori to face these situations.

First of all, we do not solve this problem by changing the used constitutive equations since this behavior is mainly due to the degradation of strength and/or to the fact that constitutive equations of geomaterials do not meet a normality condition.

Second, the situation has to be considered differently at the element level and at the structural level. The non-uniqueness of the solution is more serious at the micro-scale since it can be concluded in some cases that existence of a constitutive equation itself is questionable.

From the experimental point of view it is well known, although not often reported, that as soon as degradation is clearly observed, the experimental data are more scattered and the experiments are poorly or not at all reproducible (see [12] and [13] for instance).

Finally, the non-uniqueness of the theoretical and numerical solutions is observed when experimental results are also questionable, which is an important clue to think that this is neither a deficiency of the modeling nor of the experimental results. We have to search solutions to deal with this unavoidable situation. Some of them can certainly be outside of the scope of the classical continuum mechanics framework.

REFERENCES

[1] Chambon, R., Caillerie, D.: Existence and uniqueness for boundary value problem involving incrementally non linear models. Int. J. Solids Struct. 36, 5089–5099 (1999)

[2] Caillerie, D., Chambon, R.: Existence and uniqueness for rate problem of geomechanics. Revue Française de Génie Civil. 8, 537–561 (2004)

[3] Rice, J.: The localisation of plastic deformation. In: Proceeding of I.C.T.A.M., pp. 207–220. North Holland Publishing, Amsterdam (1976)

[4] Nova, R.: Controllability and incremental response. J. Mech. Behav. Materials 5, 221–243 (1994)

[5] Chambon, R.: Some theoretical results about second order work, uniqueness, existence and controllability independent of the constitutive equation. J. Eng. Mathematics 52, 53–61 (1994)

[6] Chambon, R., Crochepeyre, S., Desrues, J.: Localization criteria for non-linear constitutive equations of geomaterials. Mechanics of Cohesive-Frictional Materials 5, 61–821 (2000)

[7] Chambon, R., Crochepeyre, S., Charlier, R.: An algorithm and a method to search bifurcation points in non linear problems. Int. J. Num. Meth. Eng. 51, 315–332 (2001)

[8] Chambon, R., Moullet, J.-C.: Uniqueness studies in boundary value problems involving some second gradient models. Comp. Meth. App. Mech. and Eng. 193, 2771–2796 (2004)

[9] Besuelle, P., Chambon, R., Collin, F.: Switching mode of deformation in post localization solutions with a quasi brittle material. J. Mech. Mater. Struct. 1, 1115–1134 (2006)

[10] Sieffert, Y., AlHolo, S., Chambon, R.: Loss of uniqueness of numerical solutions of the borehole problem modeled with enhanced media. Int. J. Solids Struct. 46, 3173–3197 (2009)

[11] Bilbie, G., Dascalu, C., Chambon, R., Caillerie, D.: Microfracture instabilities in granular solids. Acta Geotechnica 3, 25–35 (2009)

[12] Desrues, J.: La localisation de la déformation dans les matériaux granulaires. Ph D Thesis Grenoble University (1984)

[13] Desrues, J., Viggiani, G.: Strain localization in sand: an overview of the experimental results obtained in grenoble using stereophotogrammetry. Int. J. Numer. Anal. Meth. Geomech. 28, 279–321 (2004)

COUPLING DISCRETE ELEMENTS AND MICROPOLAR CONTINUUM THROUGH AN OVERLAPPING REGION IN ONE DIMENSION

Richard A. Regueiro[1,*] and Beichuan Yan[2]

[1] Department of Civil, Environmental, and Architectural Engineering
University of Colorado at Boulder
1111 Engineering Dr., Boulder, CO 80309
*e-mail: regueiro@colorado.edu
[2] Community Surface Dynamics Modeling System
University of Colorado at Boulder
3100 Marine St., Boulder, CO 80303
e-mail: beichuan.yan@colorado.edu, csdms.colorado.edu

Summary. The extended abstract provides the initial derivation of the coupling via an overlapped region of elastic glued circular disk discrete elements with simplified interparticle elastic constitutive relations, and a micropolar continuum finite element formulation reduced by Timoshenko beam kinematics including axial stretch. This one-dimensional form of the model provides a simple model problem from which to investigate the coupling between axial, transverse, and rotational degrees of a discrete disk representation and a micropolar continuum model of a string of elastic glued circular grains.

Keywords: discrete elements, micropolar continuum finite elements, overlap coupling.

1 INTRODUCTION

Interfacial mechanics between granular materials and deformable solid bodies involve large shear deformation and grain motion at the interface. To resolve such granular physics at the grain scale in contact with the deformable solid, but in a computationally tractable manner, a concurrent multiscale computational method is needed [1].

As a simple problem to verify the method, a one-dimensional (1D) string of glued discrete elements is overlapped with a linear elastic micropolar continuum finite element implementation. The overlap coupling is enabled by the bridging scale decomposition method [2, 3], but now with rotational degrees of freedom (dof) in addition to axial and transverse displacements. The abstract presents the initial formulation of the 1D elastic glued discrete elements and linear elastic micropolar continuum reduced by Timoshenko beam kinematics with axial stretch.

2 DISCRETE ELEMENT AND MICROPOLAR CONTINUUM FINITE ELEMENT FORMULATIONS IN ONE DIMENSION

2.1 Discrete Element Model of Glued Elastic Circular Disks

Since the disk particles are assumed glued and elastic, we simplify their interparticle constitutive behavior by linear elastic springs in axial, transverse, and rotational dof.

R.I. Borja (Ed.): Multiscale and Multiphysics Processes in Geomechanics, SSGG, pp. 41–44.
springerlink.com

For eventual extension to frictional sliding, we can incorporate later the realistic particle geometry in three dimensions, with particle contact search and interparticle friction and Hertzian contact [4].

The balance of linear and angular momentum for a system of elastic particles in contact may be written as

$$\boldsymbol{M}^Q\ddot{\boldsymbol{Q}} + \boldsymbol{F}^{INT,Q}(\boldsymbol{Q}) = \boldsymbol{F}^{EXT,Q} \tag{1}$$

$$\boldsymbol{M}^Q = \mathop{\mathbf{A}}_{\delta=1}^{N} \boldsymbol{m}_\delta^Q \,; \quad \boldsymbol{m}_\delta^Q = \begin{bmatrix} m_\delta & 0 & 0 \\ 0 & m_\delta & 0 \\ 0 & 0 & m_\delta^\omega \end{bmatrix}$$

$$\boldsymbol{F}^{INT,Q} = \mathop{\mathbf{A}}_{\delta=1}^{N} \boldsymbol{f}_\delta^{INT,Q} \,; \quad \boldsymbol{f}_\delta^{INT,Q} = \sum_{\epsilon=1}^{n_c} \begin{bmatrix} \boldsymbol{f}^{\epsilon,\delta} \\ \boldsymbol{f}_\omega^{\epsilon,\delta} \end{bmatrix}$$

$$\boldsymbol{F}^{EXT,Q} = \mathop{\mathbf{A}}_{\delta=1}^{N} \boldsymbol{f}_\delta^{EXT,Q} \,; \quad \boldsymbol{f}_\delta^{EXT,Q} = \begin{bmatrix} \boldsymbol{f}^{EXT,\delta} \\ \boldsymbol{\ell}^{EXT,\delta} \end{bmatrix}$$

where $\boldsymbol{M}^Q$ is the mass and rotary inertia matrix for a system of N particles, $\boldsymbol{m}_\delta^Q$ is the mass and rotary inertia matrix for particle δ, m_δ is the mass for particle δ in x_1 and x_2 directions, m_δ^ω is the rotary inertia for particle δ about the x_3 axis, $\mathbf{A}_{\delta=1}^{N}$ is an assembly operator to obtain the system matrices from the individual particle matrices and contact vectors, $\boldsymbol{F}^{INT,Q}$ the internal force and moment vector associated with n_c particle contacts which will be a linear function of particle displacements and rotations when particles translate and rotate, $\boldsymbol{f}_\delta^{INT,Q}$ the resultant internal force and moment vector for particle δ, $\boldsymbol{f}^{\epsilon,\delta}$ the internal force vector for particle δ at contact ϵ, $\boldsymbol{f}_\omega^{\epsilon,\delta}$ the internal moment vector at the centroid of particle δ, $\boldsymbol{F}^{EXT,Q}$ the assembled external force and moment vector, $\boldsymbol{f}_\delta^{EXT,Q}$ the external body force and moment vector for particle δ, $\boldsymbol{f}^{EXT,\delta}$ the external body force vector at the centroid of particle δ, and $\boldsymbol{\ell}^{EXT,\delta}$ the external body moment vector at the centroid of particle δ. $\boldsymbol{Q}$ is the generalized dof vector for particle displacements and rotations

$$\boldsymbol{Q} = [\boldsymbol{q}_\delta, \boldsymbol{q}_\epsilon, \ldots, \boldsymbol{q}_\eta, \boldsymbol{\omega}_\delta, \boldsymbol{\omega}_\epsilon, \ldots, \boldsymbol{\omega}_\eta]^T \,, \quad \delta, \epsilon, \ldots, \eta \in \mathcal{A} \tag{2}$$

where $\boldsymbol{q}_\delta$ is the displacement vector of particle δ, $\boldsymbol{\omega}_\delta$ its rotation vector, and $\mathcal{A}$ is the set of free particles (see Fig.1). In general, a superscript Q denotes a variable associated with particle motion, whereas a superscript D will denote a variable associated with continuum deformation.

2.2 Finite Element Model of Elastic Micropolar Continuum with Reduced Kinematics

Following the formulation of Eringen [5], the balance equations for linear and angular momentum ignoring body forces and couples may be written as

$$\sigma_{lk,l} - \rho\dot{v}_k = 0 \tag{3}$$

$$m_{lk,l} + e_{kmn}\sigma_{mn} - \rho\dot{\beta}_k = 0 \tag{4}$$

where σ_{lk} is the unsymmetric Cauchy stress tensor, ρ the mass density, $v_k = \dot{u}_k$ the velocity vector, m_{lk} the unsymmetric couple stress, e_{kmn} the permutation operator,

$\beta_k = j\dot{\varphi}_k$ the intrinsic spin per unit mass, j the micro-inertia, indices $k, l, \cdots = 1, 2, 3$, and $(\bullet)_{,l} = \partial(\bullet)/\partial x_l$ denotes partial differentiation with respect to the spatial coordinate x_l. Introducing w_k and η_k as weighting functions for the macro-displacement vector u_k and micro-rotation vector φ_k, respectively, we apply the Method of Weighted Residuals to formulate the partial differential equations in (3) and (4) into weak form. The weak, or variational, equations then result as

$$\int_{\mathcal{B}} \rho w_k \dot{v}_k dv + \int_{\mathcal{B}} w_{k,l}\sigma_{lk} dv = \int_{\Gamma_t} w_k t_k da \tag{5}$$

$$\int_{\mathcal{B}} \rho \eta_k \dot{\beta}_k dv + \int_{\mathcal{B}} \eta_{k,l} m_{lk} dv - \int_{\mathcal{B}} \eta_k e_{kmn}\sigma_{mn} dv = \int_{\Gamma_r} \eta_k r_k da \tag{6}$$

where $\mathcal{B}$ is the volume of the continuum body, $t_k = \sigma_{lk} n_l$ is the applied traction on the portion of the boundary Γ_t with outward normal vector n_l, and $r_k = m_{lk} n_l$ is the applied surface couple on the portion of the boundary Γ_r.

We introduce the following reduced kinematics for Timoshenko beam with stretch [6] as

$$\boldsymbol{u} = \begin{bmatrix} u_1 \\ u_2 \\ u_3 \end{bmatrix} = \begin{bmatrix} u - x_2\theta \\ v \\ w \end{bmatrix}, \; \boldsymbol{\varphi} = \begin{bmatrix} 0 \\ 0 \\ \theta \end{bmatrix} \tag{7}$$

where u is the axial displacement, v the transverse displacement, and θ the rotation. (Note that the third component of φ is a rotation about the x_3 axis, thus a rotation in the $x_1 - x_2$ plane.) We ignore the transverse displacement and stretch w in the x_3 direction, although we can solve for σ_{33} if need be (i.e., when we extend to pressure-sensitive plasticity for the micropolar or micromorphic continuum [7]). After substituting into the coupled weak form, assuming a mixed 1D finite element (quadratic in v, and linear in u and θ), details omitted, we arrive at the following coupled finite element matrix form as

$$\boldsymbol{M}^D \ddot{\boldsymbol{D}} + \boldsymbol{K}^D \boldsymbol{D} = \boldsymbol{F}^{EXT,D} \tag{8}$$

$$\boldsymbol{M}^D = \begin{bmatrix} \boldsymbol{M}^{uu} & \boldsymbol{0} & -\boldsymbol{M}^{u\theta} \\ \boldsymbol{0} & \boldsymbol{M}^{vv} & \boldsymbol{0} \\ -\boldsymbol{M}^{\theta u} & \boldsymbol{0} & \boldsymbol{M}^{\theta\theta} \end{bmatrix} \quad \boldsymbol{D} = \begin{bmatrix} \boldsymbol{d}_u \\ \boldsymbol{d}_v \\ \boldsymbol{d}_\theta \end{bmatrix}$$

$$\boldsymbol{K}^D = \begin{bmatrix} \boldsymbol{K}^{uu} & \boldsymbol{0} & -\boldsymbol{K}^{u\theta} \\ \boldsymbol{0} & \boldsymbol{K}^{vv} & -\boldsymbol{K}^{v\theta} \\ -\boldsymbol{K}^{\theta u} & -\boldsymbol{K}^{\theta v} & \boldsymbol{K}^{\theta\theta} \end{bmatrix} \quad \boldsymbol{F}^{EXT,D} = \begin{bmatrix} \boldsymbol{F}_F \\ \boldsymbol{F}_V \\ \boldsymbol{F}_M + \boldsymbol{F}_{M^\theta} \end{bmatrix} \tag{9}$$

where $\boldsymbol{M}^D$ is symmetric, but $\boldsymbol{K}^D$ is unsymmetric. The submatrices of $\boldsymbol{M}^D$ and $\boldsymbol{K}^D$ couple u, v, and θ.

3 COUPLING APPROACH

The coupling implementation extends to particle mechanics and micropolar continuum [1] the "bridging scale decomposition" proposed by Wagner and Liu [2] and modifications thereof by Klein and Zimmerman [3]. The overall approach is illustrated in Fig.1, where the particle region in the lower left corner would be interfacing a deformable solid body. The purple background denotes the finite element (FE) overlap region $\tilde{\mathcal{B}}^h$ with underlying ghost particles, aqua blue the FE continuum region $\bar{\mathcal{B}}^h$ with no underlying particles, and white background (with brown particles) the free particle region $\hat{\mathcal{B}}^h \cup \mathcal{B}^{DE}$. Details are discussed in [1], and 1D illustration will be presented at the workshop and published in a future paper.

4 CONCLUSIONS

The abstract presented the preliminaries of coupling DE regions and micropolar FE regions for eventual simulation of granular soil-tire/tool/ penetrometer applications, wherein three-dimensional DE formulation and finite strain micromorphic continuum FE are required [7].

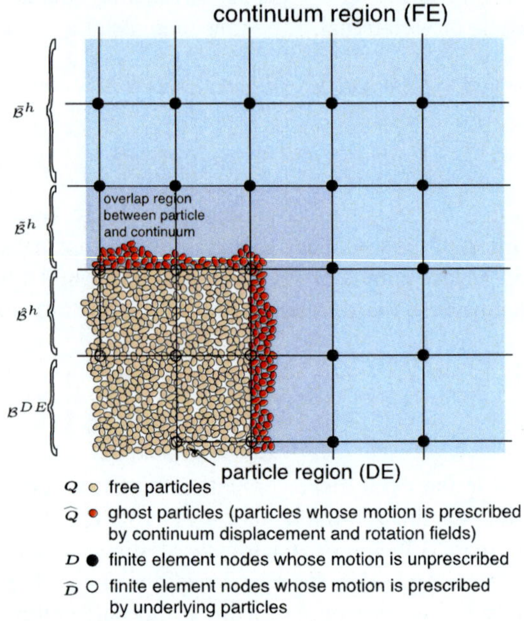

Fig. 1. 2D illustration of coupling between particle and continuum regions of a cohesionless granular material.

ACKNOWLEDGEMENTS

Funding for this research was provided by National Science Foundation grant CMMI-0700648, Army Research Office grant W911NF-09-1-0111, and the Army Research Laboratory. This funding is gratefully acknowledged.

REFERENCES

[1] Regueiro, R.A., Yan, B.: Concurrent multiscale computational modeling for dense dry granular materials interfacing deformable solid. In: Wan, R., Alsaleh, M., Labuz, J. (eds.) Bifurcations, Instabilities and Degradations in Geomaterials, pp. 251–273. Springer, Berlin (2011)

[2] Wagner, G.J., Liu, W.K.: Coupling of atomistic and continuum simulations using a bridging scale decomposition. J. Comput. Phys. 190(1), 249–274 (2003)

[3] Klein, P.A., Zimmerman, J.A.: Coupled atomistic-continuum simulations using arbitrary overlapping domains. J. Comput. Phys. 213(1), 86–116 (2006)

[4] Yan, B., Regueiro, R.A., Sture, S.: Three dimensional discrete element modeling of granular materials and its coupling with finite element facets. Eng. Comput. 27(4), 519–550 (2010)

[5] Eringen, A.C.: Theory of Micropolar Elasticity. In: Liebowitz, H. (ed.) Fracture, An Advanced Treatise, pp. 622–729. Academic Press, New York (1968)

[6] Pinsky, P.M.: Finite Element Structural Analysis, course notes, Stanford University

[7] Regueiro, R.A.: Finite strain micromorphic pressure-sensitive plasticity. J. Eng. Mech. 135, 178–191 (2009)

MULTISCALE MODELING OF GRANULAR MATTER: A HIERARCHICAL SCHEME

José E. Andrade[1,*] and Carlos F. Avila[2]

[1] Division of Engineering & Applied Science
California Institute of Technology
Pasadena, CA 91125
e-mail: jandrade@caltech.edu
[2] Division of Engineering & Applied Science
California Institute of Technology
Pasadena, CA 91125
e-mail: cavila@caltech.edu

Summary. This paper briefly describes a hierarchical multiscale method used to model the behavior of granular matter by linking the fundamental granular scale with the continuum scale, which is being modeled using frictional elasto-plasticity. We show a particular example where grain kinematics are extracted from discrete element calculations and successfully coupled with finite element procedures, without resorting to any hardening evolution functions. The accuracy of the proposed method is verified via a numerical example.

Keywords: granular, microstructure, multiscale, plasticity.

1 INTRODUCTION

Continuum plasticity models have a definite place in science and engineering due to their versatility and ability to capture average material behavior relatively accurately. However, one major problem with phenomenologically based plasticity models for granular materials is their inability to evolve in situations where conditions deviate significantly from homogeneous deformations (e.g., within deformation bands). On the other hand, discrete models (e.g., DEM) are limited due to their computational expense and have not been able to predict complex material behavior in granular matter at lab scale or larger. It is, therefore, apparent that there is a need for multiscale models that can link discrete mechanics with continuum mechanics and exploit the advantages of both frameworks. In this work, we present some results based on a recently developed hierarchical multiscale scheme for granular materials [1]. The central idea used in the multiscale scheme is the calculation of plastic internal variables directly from the micromechanical process. We focus on pure numerical simulations based on the discrete element method (DEM) and the finite element method (FEM) and on verifying the proposed techinique. Validation with actual experiments is conducted elsewhere [1]. However, verification is a crucial step in demonstrating the ability of the multiscale scheme in capturing the essence of the material behavior encoded at the grain scale.

R.I. Borja (Ed.): Multiscale and Multiphysics Processes in Geomechanics, SSGG, pp. 45–48.
springerlink.com

2 MULTISCALE GRANULAR PLASTICITY

Consider the canonical elasto-plastic continuum tangent operator $\boldsymbol{c}^{\mathrm{ep}}$—relating the change in total strain $\dot{\boldsymbol{\epsilon}}$ to an equivalent change in stress $\dot{\boldsymbol{\sigma}} = \boldsymbol{c}^{\mathrm{ep}} : \dot{\boldsymbol{\epsilon}}$, i.e.,

$$\boldsymbol{c}^{\mathrm{ep}} = \boldsymbol{c}^{\mathrm{e}} - \frac{1}{\chi}\boldsymbol{c}^{\mathrm{e}} : \boldsymbol{q} \otimes \boldsymbol{f} : \boldsymbol{c}^{\mathrm{e}}; \qquad \chi = \boldsymbol{q} : \boldsymbol{c}^{\mathrm{e}} : \boldsymbol{f} \tag{1}$$

with $\boldsymbol{c}^{\mathrm{e}}$ as the (linear) elastic tangent, $\boldsymbol{f}$ and $\boldsymbol{q}$ as the gradients (with respect to the stress) to the yield surface F and plastic potential Q, respectively. As shown in Figure 1, in the case of frictional materials, the yield function and plastic potential are functions of the stress and plastic internal variables (PIVs) so that $F = F(\boldsymbol{\sigma}, \mu)$ and $Q = Q(\boldsymbol{\sigma}, \beta)$. In granular materials, μ and β typically represent physical variables, such as, friction and dilatancy, respectively. This physical meaning will enable direct extraction of μ and β from the microstructure and will render phenomenological evolution laws unnecessary.

Figure 1 shows the essence of the hierarchical scheme proposed in this paper. The central idea is to use the underlying granular structure of the material, modeled in this case by discrete element method, to extract the evolution of the PIVs required by the continuum plasticity model. In doing so, the plasticity evolution is reconstructed by effectively linking the granular microstructure with the continuum representation. Within the context of granular plasticity, the PIVs are friction μ and dilatancy β, which can both be extracted based on the micromechanical state: stress $\bar{\boldsymbol{\sigma}}$ and strain $\bar{\boldsymbol{\epsilon}}$ (See [1–3] for specific formulae to calculate micro-stress and strain).

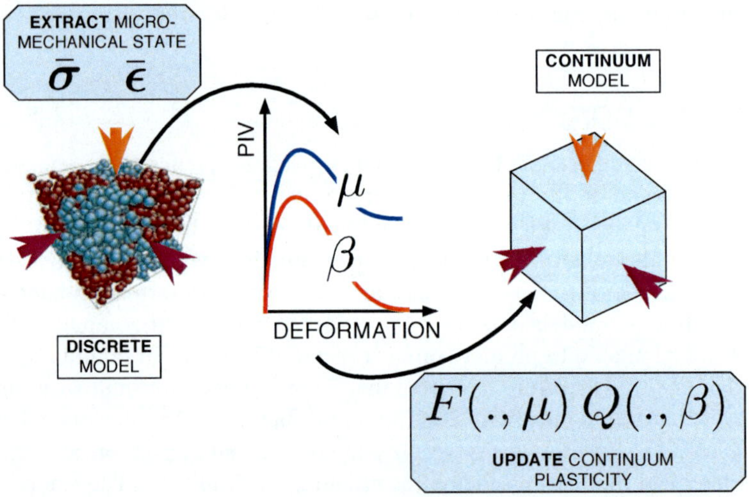

Fig. 1. Schematic of the proposed hierarchical multiscale method. Plastic internal variables are extracted from the discrete model by calculating the micro-mechanical state encoded in the micro-stress $\bar{\boldsymbol{\sigma}}$ and micro-strain $\bar{\boldsymbol{\epsilon}}$. Then, the micro-inspired evolution of the PIVs is used sequentially within a continuum plasticity calculation to predict the material behavior under equivalent boundary-value problems.

Once the micromechanical state is obtained from the discrete model, the continuum PIVs are calculated by invoking compatibility such that

$$\mu \approx -\frac{\bar{q}}{\bar{p}} \tag{2}$$

with $\bar{p} = 1/3 \operatorname{tr} \bar{\boldsymbol{\sigma}}$ as the mean normal stress invariant and $\bar{q} = \sqrt{3/2}\|\operatorname{dev} \bar{\boldsymbol{\sigma}}\|$ as the deviatoric stress invariant. Similarly, (plastic) dilatancy can be extracted by assuming that the incremental elastic strains are negligeable, and hence,

$$\beta \approx \frac{\Delta \bar{\epsilon}_{\mathrm{v}}}{\Delta \bar{\epsilon}_{\mathrm{s}}} \tag{3}$$

where Δ implies change, $\bar{\epsilon}_{\mathrm{v}} = \operatorname{tr} \bar{\boldsymbol{\epsilon}}$, and $\bar{\epsilon}_{\mathrm{s}} = \sqrt{2/3}\|\operatorname{dev} \bar{\boldsymbol{\epsilon}}\|$.

The micromechanically based evolution for μ and β is obtained as a function of the deformation process (e.g., deviatoric invariant) and then subsequently used to model the behavior of the material according to the elastoplastic continuum tangent given in equation (1). A numerical example is included in the next section to verify the accuracy of the hierarchical scheme shown in Figure 1.

3 NUMERICAL EXAMPLE

The granular assembly shown on the left of Figure 1 was used to perform both direct numerical simulations (DNS) and multiscale calculations to verify the proposed multiscale scheme. The assembly is composed of 1,800 poly-dispersed particles and divided into 8 quadrants represented by the different colors in Figure 1. Division into subregions allows the procedure to exploit parallelization. As for boundary conditions,

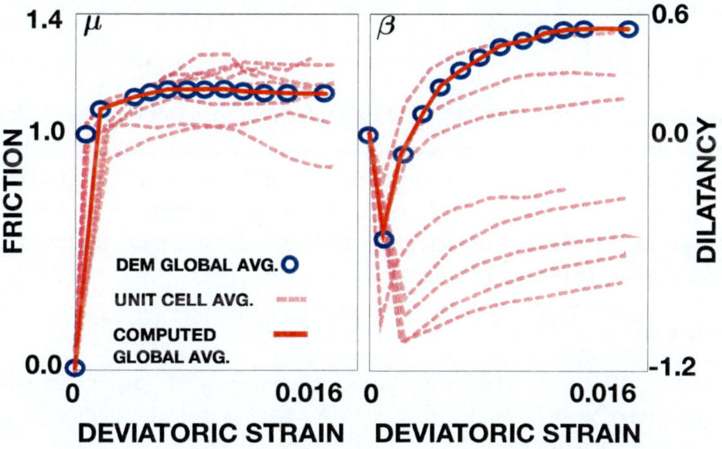

Fig. 2. PIV evolution from DEM computations (direct numerical simulations). Friction and dilatancy are extracted from each sub-region (pink curves) and then used to compute a global average (red curve). Additionally, the global stress and strain are used to calculate the global evolution of the PIVs stemming from the entire sample in the DNS (blue points).

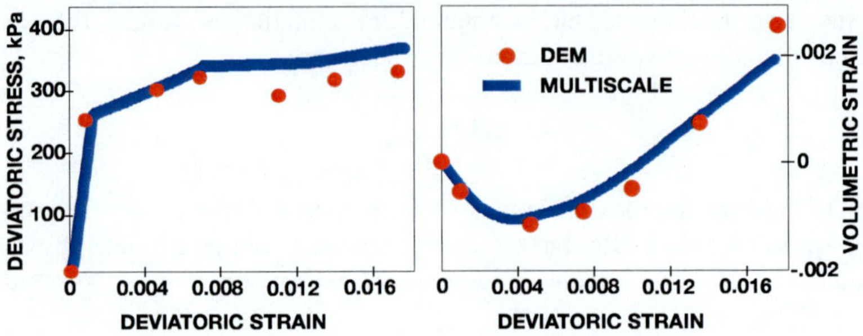

Fig. 3. Stress-strain response comparing DNS and multiscale calculations.

the granular assembly is confined laterally using a confinement pressure of 170 kPa. The sample is sheared quasi-statically with an axial strain rate of 2E-4 sec^{-1}. Using equations (2) and (3), the micromechanically-based evolution of μ and β was obtained for each of the sub-regions in the assembly. Figure 2 shows the evolution of the PIVs for each of the 8 sub-regions in the granular assembly and compares the average response with that calculated from the global (boundary) measures of stress and strain.

Using the evolution curves for μ and β, the behavior of the material is simulated using a finite element mesh composed of 8 brick elements (each corresponding to the quadrants described above). The continuum model is run using the plasticity model presented in the previous section and with exactly the same boundary conditions as those imposed on the DNS. The global stress–strain curves obtained are shown in Figure 3. It is apparent from these results that the proposed hierarchical multiscale technique can extract the global material behavior implied at the granular scale and that these calculations open the door to more accurate multiscale modeling of granular materials in the near future.

4 CONCLUSIONS

We have shown a hierarchical multiscale technique to model the behavior of granular materials under shear. The procedure is based on extracting key plastic internal variables from the micromechanics and then passing them to the continuum plasticity model. A verification example was shown with excellent comparisons achieved for the global stress–strain response. This opens the door to more accurate multiscale modeling of granular materials in the near future.

REFERENCES

[1] Andrade, J.E., Avila, C.F., Hall, S.A., Lenoir, N., Viggiani, G.: Multiscale modeling and characterization of granular matter: from grain kinematics to continuum mechanics. Journal of the Mechanics and Physics of Solids 59, 237–250 (2011)
[2] Andrade, J.E., Tu, X.: Multiscale framework for behavior prediction in granular media. Mechanics of Materials 41, 652–669 (2009)
[3] Wellmann, C., Lillie, C., Wriggers, P.: Homogenization of granular material modeled by a three-dimensional discrete element method. Computers and Geotechnics 35, 394–405 (2007)

STRENGTH OF GEOMATERIALS: MULTISCALE THEORIES AND EXPERIMENTS AT APPROPRIATE PROBLEM-DEPENDENT LENGTH SCALES

Bernhard Pichler, Christian Hellmich, and Josef Eberhardsteiner

Institute for Mechanics of Materials and Structures
Vienna University of Technology (TU Wien)
Karlsplatz 13/202, A-1040 Vienna, Austria
e-mail: Bernhard.Pichler@tuwien.ac.at, Christian.Hellmich@tuwien.ac.at,
Josef.Eberhardsteiner@tuwien.ac.at, web page: http://www.imws.tuwien.ac.at

Summary. Quantification of the strength of geomaterials is often the key to effective solutions to problems in geoengineering. The scales where underlying processes are favorably studied and quantified may largely vary. We here consider two quite distinct cases, related to penetration resistance of gravel under rockfall, and to chemically driven strength growth in sprayed concrete (shotcrete), together with corresponding large-scale applications: rockfall protection of oil pipelines and tunneling according to the New Austrian Tunneling Method.

Keywords: continuum micromechanics, homogenization, separation of scales, upscaling.

1 INTRODUCTION

Quantification of the strength of geomaterials is often the key to effective solutions of problems in geoengineering. This task requires sufficient theoretical understanding of the underlying (multiscale) processes, which is nurtured by correspondingly designed experimental campaigns. Depending on the nature of the considered geomaterials, the scales where this interaction between theory, computations, and experiments takes place, may largely vary. We here consider two quite distinct cases, related to penetration resistance of gravel under rockfall, and to chemically driven strength growth in sprayed concrete (shotcrete), together with corresponding large-scale applications: rockfall protection of oil pipelines and tunneling according to the New Austrian Tunneling Method (NATM).

In the first case, complex microstructural transformation processes, including crushing of individual stones, frictional sliding between neighboring stones, and compaction of the hit material, are extremely complex to be directly modeled, especially when the resulting mathematical model should not only reflect observed phenomena purely qualitatively, but when such a model should allow for reliable prognoses of untested situations. Since our focus is (fortunately) on homogeneous gravel layers whose microstructural composition is free of macroscopic gradients, it was possible, on the basis of dimensionally homogeneous relationships known in the ballistic literature [1–3], to predict penetration depths and maximum impact forces of boulders impacting on gravel

R.I. Borja (Ed.): Multiscale and Multiphysics Processes in Geomechanics, SSGG, pp. 49–52.
springerlink.com

layers, with the involvement of only a single strength-like gravel quantity, the penetration resistance, described in Section 2.

In the second case, the microstructure and the composition of the considered cement-based geomaterial "shotcrete" undergoes huge changes along the dimensions of a tunnel shell, due to the chemical reaction between cement and water, called hydration. Hence, fields of mechanical properties (in particular strength) are not homogeneous, and no numerical strength values valid for the entire shotcrete shell can be given. As a remedy, a true multiscale model was developed in the framework of random homogenization theory or continuum micromechanics, going down to the level where "universal", i. e. hydration-independent, values for elasticity and strength can be identified, i. e. where unhydrated clinker, water, hydrates (the reaction products between water and cement), air, and aggregates can be discerned. This is described in Section 3.

2 UNDERSTANDING STRENGTH OF BOULDER-HIT GRAVEL – FOR PROTECTION OF OIL PIPELINES

Increasing rockfall activity in the European Alps raises the need for designing protection systems for Alpine infrastructure. As regards, e. g. oil pipelines, commonly a gravel overburden is used to cover and, hence, protect the structure [4]. Gravel layers (a) effectively dampen the impact by absorbing remarkable amounts of energy, and (b) they have also some potential in distributing the rather concentrated impact loads.

We apply dimensionless impact laws developed from ballistic engineers studying impacts of metallic projectiles onto concrete and soil targets [1–3], to rockfall onto gravel layers. This allows estimation of penetration depths and maximum impact forces as functions of boulder mass, size, shape, and impact velocity, once the strength-like material property "indentation resistance" of gravel is identified. The latter is defined at a scale of a few decimeters, being significantly larger than that of single stones (separation of scales). 95 % confidence interval for the indentation resistance of gravel was identified from real-scale impact experiments involving boulders with masses ranging from 10 to 18 tons and heights of fall ranging from 2 to 20 meters [5]. This was possible because the impact craters were significantly larger than the characteristic stone diameters, indicating that the separation of scales principle was fulfilled.

The identified 95 % confidence interval for the indentation resistance of gravel opens the door to semi-probabilistic design of rockfall protection layers, e. g. for buried steel pipelines [6]. For this purpose, (i) geologists have to define a design rockfall in terms of boulder mass, shape, height of fall, and mode of impact (ii) a specific burying depth of the pipeline is chosen, and (iii) several structural simulations of the hit structure are performed, based on different values of the indentation resistance of gravel taken out of the 95 % confidence interval. Thereby, the penetration process does not have to be modeled in full detail. It is sufficient to prescribe a traction distribution corresponding to the maximum impact force at a depth which is equal to the penetration depth at maximum impact force [4].

3 MULTISCALE QUANTIFICATION OF HYDRATION-DEPENDENT SHOTCRETE STRENGTH – FOR SAFETY ASSESSMENT OF NATM TUNNELS

Cement paste is the binder of cementitious materials. Produced by mixing a cement clinker powder with water, the material is initially a viscous suspension. The chemical reaction between clinker grains and water results in the creation of hydration products establishing a connected network of solid particles such that cement paste transforms into a solid material. The hydration process is commonly parametrized by the hydration degree, which is zero at the beginning and which reaches the value of one once all available clinker has hydrated. Since the chemical reaction is essentially temperature-dependent, macroscopic properties of cement paste are a function of hydration degree rather than a function of time.

Since the early 2000's, it is well known that continuum micromechanics-based models are able to reliably predict stiffness evolutions of both cement paste and concrete, see, e.g., [7, 8]. In this context, quasi-homogeneous material properties are defined on representative volume elements (RVEs) containing micro-heterogeneous matter. This requires the validity of the separation of scales principle, stating that the characteristic size of an RVE has to be significantly larger than the characteristic size of the micro-heterogeneity, and at the same time, significantly smaller than the structure built up by the material defined on the RVE.

Rather recently, it was shown that continuum micromechanics models can be extended towards prediction of uniaxial macroscopic compressive strength of cementitious materials [9, 10], in the framework of elasto-brittle analysis. These models assume (i) that an RVE of a cement-based material behaves linear elastic as long as deviatoric stress peaks in micron-sized needle-shaped hydrate agglomerates are smaller than the deviatoric hydrate strength, and (ii) that the macroscopic ultimate load of the material is reached, once the aforementioned deviatoric stress peaks reach the hydrate strength. Based on quadratic deviatoric strain averages [11], hydrate stress peaks are derived from the hierarchical organization of cementitious materials: at a scale from 15 to 20 microns, a polycrystalline arrangement of needle-shaped hydrate agglomerates with isotropic spatial orientations as well as spherical water and air pores is homogenized, delivering elastic and strength properties of a hydrate foam [12]. The latter represents — at a scale of half a millimeter (cement paste level) — a matrix in which spherical clinker grains are embedded. Finally, at a scale of half a decimeter (mortar/shotcrete level), cement paste forms a matrix in which spherical sand grains and aggregates are embedded.

Continuum micromechanics-based material models for shotcrete elasticity, strength, and creep [13] give access to relationships between hydration degree and mechanical properties for any shotcrete composition of interest, defined in terms of water-cement ratio and aggregate-cement ratio. Among other applications, this is beneficial for day-to-day assessments of the safety of shotcrete tunnel shells during their construction [8, 10, 14].

REFERENCES

[1] Forrestal, M.J., Altman, B.S., Cargile, S.J., Hanchak, S.J.: An empirical equation for penetration depth of ogive-nose projectiles into concrete targets. International Journal of Impact Engineering 15(4), 395–405 (1994)

[2] Forrestal, M.J., Frew, D.J., Hanchak, S.J., Brar, N.S.: Penetration of grout and concrete targets with ogive-nose steel projectiles. International Journal of Impact Engineering 18(5), 465–476 (1996)

[3] Li, Q.M., Chen, X.W.: Dimensionless formulae for penetration depth of concrete target impacted by a non-deformable projectile. International Journal of Impact Engineering 28(1), 93–116 (2003)

[4] Pichler, B., Hellmich, C., Mang, H.A., Eberhardsteiner, J.: Loading of a gravel-buried steel pipe subjected to rockfall. Journal of Geotechnical and Geoenvironmental Engineering 132(11), 1465–1473 (2006)

[5] Pichler, B., Hellmich, C., Mang, H.A.: Impact of rocks onto gravel design and evaluation of experiments. International Journal of Impact Engineering 31(5), 559–578 (2005)

[6] Pichler, B., Hellmich, C., Eberhardsteiner, J., Mang, H.A.: Semi-probabilistic design of rockfall protection layers. Computational Mechanics 42(2), 327–336 (2008)

[7] Bernard, O., Ulm, F.-J., Lemarchand, E.: A multiscale micromechanics-hydration model for the early-age elastic properties of cement-based materials. Cement and Concrete Research 33(9), 1293–1309 (2003)

[8] Hellmich, C., Mang, H.A.: Shotcrete elasticity revisited in the framework of continuum micromechanics: From submicron to meter level. Journal of Materials in Civil Engineering (ASCE) 17(3), 246–256 (2005)

[9] Pichler, B., Hellmich, C., Eberhardsteiner, J.: Spherical and acicular representation of hydrates in a micromechanical model for cement paste: Prediction of early-age elasticity and strength. Acta Mechanica 203(3-4), 137–162 (2009)

[10] Pichler, B., Scheiner, S., Hellmich, C.: From micron-sized needle-shaped hydrates to meter-sized shotcrete tunnel shells: Micromechanical upscaling of stiffness and strength of hydrating shotcrete. Acta Geotechnica 3(4), 273–294 (2008)

[11] Dormieux, L., Molinari, A., Kondo, D.: Micromechanical approach to the behavior of poroelastic materials. Journal of Mechanics and Physics of Solids 50(10), 2203–2231 (2002)

[12] Pichler, B., Hellmich, C.: Upscaling quasi-brittle strength of cement paste and mortar: A multi-scale engineering mechanics model. Cement and Concrete Research (2011), doi:10.1016/j.cemconres.2011.01.010

[13] Scheiner, S., Hellmich, C.: Continuum microviscoelasticity model for aging basic creep of early-age concrete. Journal of Engineering Mechanics (ASCE) 135(4), 307–323 (2009)

[14] Ullah, S., Pichler, B., Scheiner, S., Hellmich, C.: Improved micromechanics-based safety assessment of shotcrete tunnels, through shell-specific interpolation of measured 3D displacements. Computer Modeling in Engineering & Sciences 57(3), 279–316 (2010)

A TWO-SCALE DAMAGE LAW FOR CREEPING ROCKS

Cristian Dascalu[1,*] and Bertrand François[1,2]

[1] Laboratoire Sols Solides Structures - Risques, UJF, INPG, CNRS UMR 5521
Domaine Universitaire, B.P. 53, 38041 Grenoble cedex 9, France
*e-mail: cristian.dascalu@hmg.inpg.fr
[2] Departement ArGEnCo, Geo³- Géomécanique et Géologie de l'ingénieur
Université de Liège, FRS-FNRS Fonds de la Recherche Scientifique
Chemin des Chevreuils, 1 4000 Liège 1 Belgium
e-mail: bertrand.francois@ulg.ac.be

Summary. The theoretical developments and the numerical applications of a time-dependent damage law are presented. This law is deduced from considerations at the micro-scale where non-planar growth of micro-cracks, following a subcritical propagation criterion, is assumed. The passage from micro-scale to macro-scale is done through an asymptotic homogenization approach. Results of numerical simulations of time-dependent damage behavior are presented.

Keywords: crack rotation, homogenization, subcritical propagation, time-dependent damage.

1 INTRODUCTION

The evolution of damage in rocks is often time-dependent. At the micro-scale, such materials contain micro-cracks that evolve on non-planar paths in a rate-dependent manner, and this complex propagation strongly affects the macroscopic mechanical behavior. We assume that the micro-crack distribution may be locally approximated by a periodic one that is characterized by a micro-structural length — the distance between two adjacent micro-cracks. The source of time-dependency is the subcritical propagation of micro-cracks. This paper combines concepts of asymptotic homogenization [1,2] and subcritical and mixed crack propagation to deduce a time-dependent damage model.

2 MATHEMATICAL FORMULATION

Consider a two-dimensional isotropic elastic medium containing a locally periodic distribution of micro-cracks. Each crack is straight with a length $2a$ and an orientation of angle θ with respect to the x_1 direction (abscissa of the referential system considered at the macro-scale). The damage variables are d - the ratio between the crack length $2a$ and the distance between two micro-cracks ε that also represents the size of the periodicity cell (Fig. 1) :

$$d = \frac{2a}{\varepsilon} \tag{1}$$

and the angle θ. On the crack faces, traction free opening or frictionless contact conditions are assumed.

R.I. Borja (Ed.): Multiscale and Multiphysics Processes in Geomechanics, SSGG, pp. 53–56.
springerlink.com

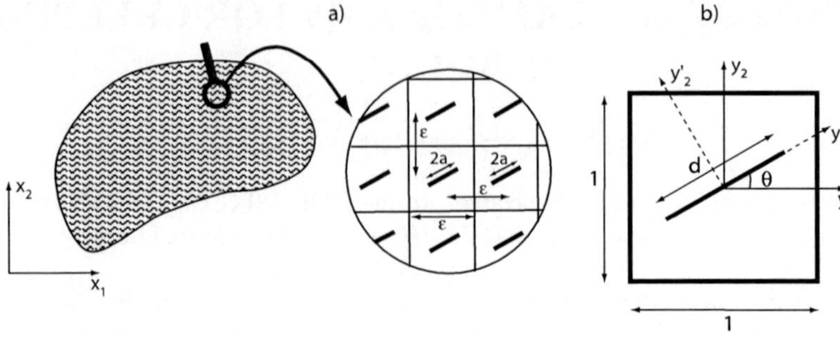

Fig. 1. (a) Fissured medium with locally periodic microstructure. (b) Unit cell with rescaled crack of length d.

2.1 Asymptotic Homogenization

The locally periodic microstructure is constructed from a unit cell $Y = [-0.5, 0.5] \times [-0.5, 0.5]$ expressed in a (y_1, y_2) orthogonal axis system centered in the middle of the crack, corresponding to the center of the cell. Then this unit cell is rescaled by the parameter ε so that the period of the material is εY. The two distinct scales are represented by the variable $\mathbf{x}$ (the macroscopic variable) and $\mathbf{y} = \mathbf{x}/\varepsilon$ (the microscopic variable). After asymptotic homogenization, we obtain the following relationships for the overall response of the material (e.g. [2,3]):

$$\frac{\partial}{\partial x_j}\Sigma_{ij}^{(0)} = 0 \quad ; \quad \Sigma_{ij}^{(0)} = C_{ijkl}e_{xkl}(\mathbf{u}^{(0)}) \quad ;$$

$$C_{ijkl}^{\pm}(d,\theta) = \frac{1}{|Y|}\int_{Y_s}(a_{ijkl} + a_{ijmn}e_{ymn}(\xi_{\pm}^{kl}))\, dy \tag{2}$$

where $\Sigma_{ij}^{(0)}$ is the macroscopic stress and C_{ijkl} the homogenized coefficients, respectively. Eq. (2) is the homogenized equation of equilibrium. $\xi_{\pm}^{pq}$ are elementary solutions of the microscopic correction $\mathbf{u}^{(1)}$ for particular $e_{xpq}(\mathbf{u}^{(0)}) = \delta_{pq}$. The distinction $\pm$ corresponds to opening (+) or contact (-) conditions of the crack lips [2]. The homogenized coefficients C_{ijkl} depend on the state of damage of the material (d and θ) and on the mechanical properties of the solid matrix (E and ν). From the integral (2), the coefficients can be initially computed for a large number of d and θ for both opening and closure states. After interpolation, polynomial expressions of $C_{ijkl}(d,\theta)$ are obtained [2,3].

2.2 Subcritical Growth of Micro-crack

The evolution of the micro-crack length is described through a subcritical criterion adapted from the Charles' law:

$$\frac{dl}{dt} = v_0\left(\frac{K_I^\star}{K_0}\right)^n \tag{3}$$

where K_0, v_0 and n are material parameters. $K_I^\star$ is the stress intensity factor for the tensile mode of rupture (Mode I) of the kinked crack (Fig. 2a). The determination of the stress intensity factors at the crack tips is made through the computation of path-independent J-, L- and M- integrals [4] for straight trajectory of micro-cracks and from

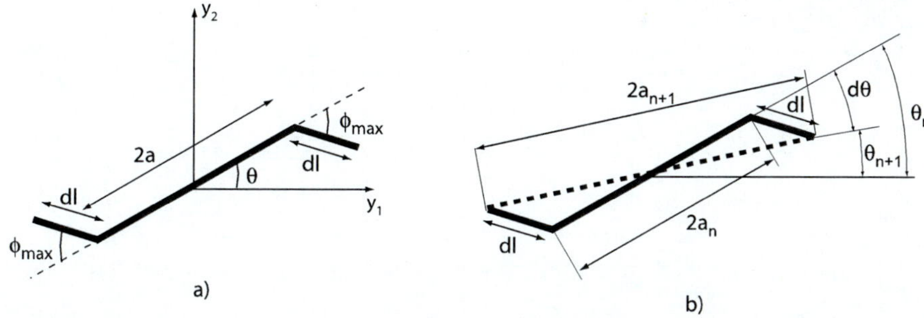

Fig. 2. Kinked crack. (a) The out-of-plane crack growth propagates in the direction that maximize the energy release rate. (b) The kinked crack (solid line) and its equivalent replacement crack (dashed line).

the polynoms given by Leblond [5] for the kinked cracks. The crack extension is assumed to propagate in the direction that maximizes the energy release rate, with a kinking angle (Fig. 2a) [6]:

$$\phi_{max} = \text{sgn}(K_{II})[0.70966\lambda^3 - 0.097725\text{sin}^2(3.9174\lambda) - 13.1588\text{tanh}(0.15199\lambda)] \; ; \quad \lambda = \frac{|K_{II}|}{K_I + |K_{II}|} \tag{4}$$

At each time increment, the kinked crack is replaced by an equivalent straight crack deduced by joining the tips of the real branched crack (Fig. 2b). We obtain in this way the evolution of damage parameters d and θ in the form of differential equations [4]:

$$\frac{dd}{dt} = \frac{2}{\varepsilon}\cos(\phi_{max})\frac{dl}{dt} \quad ; \quad \frac{d\theta}{dt} = \frac{2}{\varepsilon d}\sin(\phi_{max})\frac{dl}{dt} \tag{5}$$

3 NUMERICAL EXAMPLES

All the simulations presented in the following have been made considering biaxial loading. Plane-strain condition is considered in the third direction. Figure 3 illustrates the response of a material submitted to an uniaxial tension loading at constant vertical

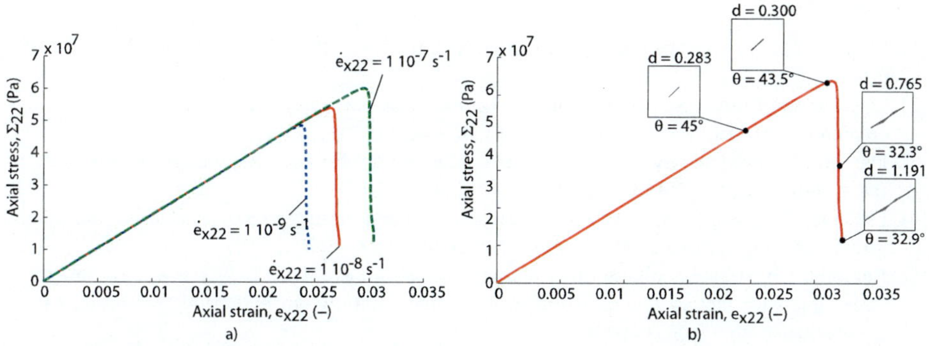

Fig. 3. Axial tension tests at various constant strain rate $\dot{e}_{x22}$. (a) The strength increases when the strain rate increases. (b) Evolution of the micro-crack in the periodic cell for $\dot{e}_{x22} = 1.10^{-8}\text{s}^{-1}$.

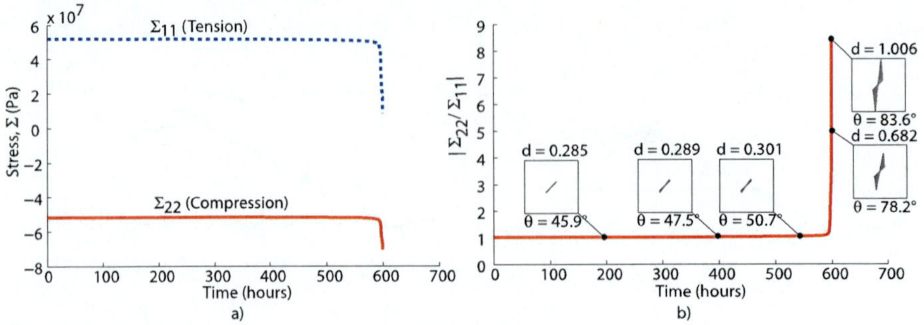

Fig. 4. Relaxation test under biaxial conditions. $e_{x22} = -0.035$ (compression) and $e_{x11} = 0.035$ (tension). Evolution with time of (a) the horizontal and vertical stresses and of (b) the ratio of anisotropy.

strain rate. Under a low strain rate, the effect of time becomes predominant and the failure appears for a lower strain level than in the case of faster loading (Figure 3a). As the damage increases, a kinked angle forms and the equivalent crack rotates, tending to be perpendicular to the principal tensile strain (Figure 3b). Figure 4a shows the evolution of horizontal and vertical stresses with time. During relaxation tests, under a biaxial combined tensile/compressive constant strain field, the subcritical micro-crack growth produces a gradual stress relaxation upon failure. As long as the crack propagates, the direction of the equivalent crack tends toward a vertical direction inducing an anisotropic response of the material (Figure 4b).

4 CONCLUSIONS

The subcritical growth of micro-cracks is responsible for the time-dependent behavior of many creeping rocks. A subcritical criterion together with a crack rotation model has been considered at the micro-scale. Upscaled equations have been obtained by an adapted asymptotic homogenization procedure. A time-dependent macroscopic damage model has been deduced. Numerical simulations in a macroscopic point of loading at constant strain rate and relaxation tests have shown the ability of the developed model to reproduce the time-dependent damage response.

REFERENCES

[1] Dascalu, C.: A two-scale damage model with material length. C.R. Mécanique 337, 645–652 (2009)
[2] Dascalu, C., Bilbie, G., Agiasofitou, E.: Damage and size effect in elastic solids: A homogenization approach. Int. J. Solid Struct. 45, 409–430 (2008)
[3] Dascalu, C., François, B., Keita, O.: A two-scale model for subcritical damage propagation. Int. J. Solid Struct. 47, 493–502 (2010)
[4] François, B., Dascalu, C.: A two-scale time-dependent damage model based on non-planar growth of micro-cracks. J. Mech. Phys. Solids 58, 1928–1946 (2010)
[5] Leblond, J.B.: Crack paths in three-dimensional elastic solids. I: two-term expansion of the stress intensity factors - application to crack path stability in hydraulic fracturing. Int. J. Solids Struct. 36, 79–103 (1999)
[6] Schütte, H., Bruhns, O.T.: On a geometrically nonlinear damage model based on a multiplicative decomposition of the deformation gradient and the propagation of microcracks. J. Mech. Phys. Solids 50, 827–853 (2002)

STRENGTH HOMOGENIZATION FOR COHESIVE-FRICTIONAL COMPOSITES: A LINEAR COMPARISON COMPOSITE APPROACH

J. Alberto Ortega*, Benjamin Gathier, and Franz-Josef Ulm

Department of Civil and Environmental Engineering
Massachusetts Institute of Technology
77 Massachusetts Avenue, Cambridge, MA, 02139, USA
*e-mail: ortega@mit.edu, ulm@mit.edu

Summary. In this work, we address the homogenization of cohesive-frictional strength properties of composite materials through the application of the Linear Comparison Composite (LCC) theory. The LCC methodology is implemented in a multiscale strength model relevant to geomaterials with heterogeneities associated with pore spaces and rigid-like grains. The proposed micromechanics solutions offer quantitative assessments of the effects of grain-scale properties and microstructure on the overall strength of the composite material.

Keywords: cohesive-frictional materials, linear comparison composite.

1 INTRODUCTION

The prediction of strength properties of geomaterials remains an active research field in view of their highly heterogeneous and compositionally diverse natures. The conventional approach to characterizing the strength behaviors is the development of empirical criteria (e.g. Mohr-Coulomb, Drucker-Prager, Hoek-Brown) that reproduce macroscopic strength-of-materials test results. However, the parameters of these models must be calibrated for the specific material under investigation. Recent developments in non-linear micromechanics have focused instead on the prediction of macroscopic strength criteria for composites based on the behaviors of the material constituents and microstructure. The micromechanics approach, originally developed for modeling purely cohesive materials (see e.g. [1]), has been extended to cohesive-frictional solids intermixed with porosity and rigid grains [2–4]. This work addresses the homogenization of cohesive-frictional strength properties of composite materials through the application of the Linear Comparison Composite (LCC) theory. The methodology is implemented in a multiscale strength model relevant to geomaterials [5,6].

2 STRENGTH HOMOGENIZATION AND LCC APPROACH

The strength homogenization problem is framed within yield design theory with the focus of determining the macroscopic dissipation capacity of a representative elementary volume (*rev*) of the composite material from the microscopic behaviors (for details, see [4]):

$$\Pi_{\mathrm{hom}}(\mathbf{D}) = \inf_{\underline{v}' \in \mathcal{V}(\mathbf{D})} \overline{\pi\left(\underline{z}, \mathbf{d}\left(\underline{v}'\right)\right)}\ ;\ \mathcal{V}(\mathbf{D}) = \left\{\underline{v}', \underline{v}'\left(\underline{z}\right) = \mathbf{D}\cdot\underline{z} \quad \left(\forall \underline{z} \in \partial\Omega\right)\right\} \quad (1)$$

R.I. Borja (Ed.): Multiscale and Multiphysics Processes in Geomechanics, SSGG, pp. 57–60.
springerlink.com

where the support function $\pi(\mathbf{d})$ denotes the maximum local dissipation capacity, and $\mathbf{D} = \overline{\mathbf{d}(\underline{v}')}$ is the macroscopic strain rate tensor corresponding to the volume average of the local strain rates linked to kinematically admissible velocity fields $\underline{v}'$. Based on the yield design definition relating stresses to the dissipation function, the stresses at the boundary of the domain of strength-compatible states delineate the homogenized yield criterion $\mathcal{F}(\mathbf{\Sigma})$:

$$\mathbf{\Sigma} \in \partial G_{\text{hom}} \Longleftrightarrow \mathbf{\Sigma} = \partial \Pi_{\text{hom}}(\mathbf{D}) / \partial \mathbf{D} \; ; \; G_{\text{hom}} = \{\mathbf{\Sigma}, \mathcal{F}(\mathbf{\Sigma}) \leq 0\} \tag{2}$$

The implementation of yield design precludes the use of linear homogenization techniques for the solution of (1). An approach to resolve the non-linear homogenization problem is the use of the Linear Comparison Composite (LCC) theory [7], which estimates the behavior of the non-linear composite in terms of a suitably chosen linear thermoelastic comparison composite with similar microstructure. To this end, the LCC approach arrives at the following stationary estimate of the macroscopic dissipation function:

$$\widetilde{\Pi}_{\text{hom}}(\mathbf{D}) = \underset{\mathbb{C},\tau}{\text{stat}}[\Psi(\mathbf{D}) + \sum f^i \Upsilon^i] \tag{3}$$

where $\Psi(\mathbf{D})$ is the strain rate energy function of the linear comparison composite (with modulus $\mathbb{C}$ and prestress τ) obtained from linear homogenization theory, and Υ expresses the difference between the local dissipation capacity and the strain rate energy, being constant in each phase of volume fraction f^i (Υ is zero in the pore space).

3 MULTISCALE STRENGTH MODEL

For the implementation of the LCC methodology, two classical problems in strength homogenization are considered: a porous solid and a solid intermixed with rigid-like inclusions. These reference configurations are incorporated into a generalized multiscale model displayed in Fig. 1. The pressure-sensitive behavior of the solid phase (level 0) is characterized by a Drucker-Prager criterion (cohesion c^s and friction α). The application of the LCC approach delivers the following dissipation function for the porous solid (level I):

$$\widetilde{\Pi}^I_{\text{hom}}(D_v, D_d) = \Sigma^I_{\text{hom}} D_v - \text{sgn}\left(2\alpha^2\mathcal{K}^I - \eta\right)\sqrt{A^I_{\text{hom}} D_v^2 + 4B^I_{\text{hom}} D_d^2} \tag{4}$$

where $D_v = \text{tr}(\mathbf{D})$, $D_d^2 = \frac{1}{2}\text{dev}(\mathbf{D}) : \text{dev}(\mathbf{D})$, and the homogenization factors are:

$$\frac{A^I_{\text{hom}}}{(c^s)^2} = \frac{\eta^2\mathcal{K}^I\left(\eta - \alpha^2\mathcal{K}^I\right)}{\left(\eta - 2\alpha^2\mathcal{K}^I\right)^2} \; ; \; \frac{B^I_{\text{hom}}}{(c^s)^2} = \frac{\eta\mathcal{M}^I\left(\eta - \alpha^2\mathcal{K}^I\right)}{\eta - 2\alpha^2\mathcal{K}^I} \; ; \; \frac{\Sigma^I_{\text{hom}}}{c^s} = \frac{\eta\alpha\mathcal{K}^I}{2\alpha^2\mathcal{K}^I - \eta} \tag{5}$$

The inclusion morphology factors $\mathcal{K}^I, \mathcal{M}^I$ are dimensionless quantities derived from classical results of linear micromechanics (refer to [4,5]), which synthesize relevant microstructural and mechanical information. These factors depend on the solid packing density $\eta = 1 - \phi_0/(1 - f^{inc})$ and the ratio of shear-to-bulk moduli of the comparison

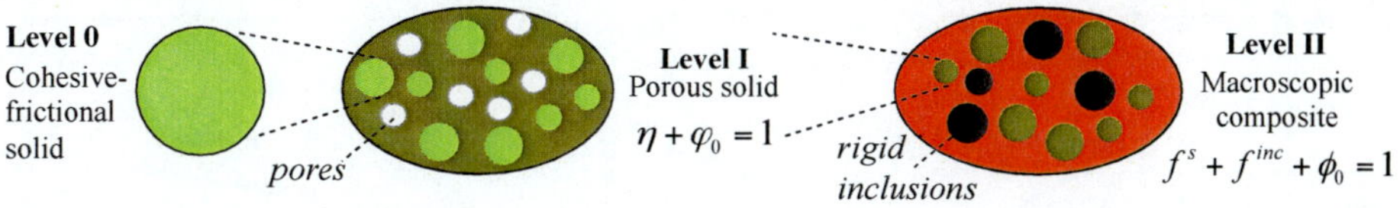

Fig. 1. Multiscale model for geomaterials, which are formed by a solid phase, inclusions, and porosity.

composite, which is explicitly resolved by the LCC method as $g^s/k^s = \alpha^2$. The application of the yield design definition (2) to (4) provides the strength criterion for the porous solid:

$$\mathcal{F}^I(\mathbf{\Sigma}) = \operatorname{sgn}\left(B^I_{\mathrm{hom}}\right)\left[\left(\Sigma_m - \Sigma^I_{\mathrm{hom}}\right)^2 / A^I_{\mathrm{hom}} + \Sigma_d^2 / B^I_{\mathrm{hom}} - 1\right] \leq 0 \qquad (6)$$

where Σ_m, Σ_d are the mean and deviatoric stresses. A similar criterion to (6) is obtained for the strength domain of the porous solid-inclusion composite (level II). The corresponding homogenization factors are detailed in [5,6] for different microstructures and interface conditions.

4 RESULTS

The LCC strength criteria are compared with the results obtained by the pioneering work of Dormieux and coworkers for benchmark microstructures. The effective strain rate approach treats the strength homogenization akin to a viscous flow problem, and employs the secant methods of non-linear homogenization. Fig. 2a displays the elliptical strength domains predicted for a porous solid with granular microstructure ($\mathcal{K}^I, \mathcal{M}^I$ modeled by the self-consistent homogenization scheme). The effective strain rate and LCC solutions compare adequately for moderate mean stress states. However, the LCC method delivers a more general solution for strength upscaling as it removes the restriction of the incompressibility of the viscous solid ($g^s/k^s \to 0$). The comparisons between the two methods for the case of a solid ($\eta \to 1$) reinforced by rigid inclusions are displayed in Fig. 2b, for which the macroscopic criterion (6) reduces to the form $\mathcal{F}^{II}(\mathbf{\Sigma}) = \Sigma_d + \alpha_{hom}(\Sigma_m - c^s/\alpha) \leq 0$. Fig. 2b shows the satisfactory comparisons for homogenized friction coefficients associated with a matrix-inclusion morphology ($\mathcal{K}^{II}, \mathcal{M}^{II}$ modeled by the Mori-Tanaka homogenization scheme) and different interface conditions.

The LCC homogenization model supports two strength domain forms: elliptical and hyperbolic (see Fig. 2a). The transition between these regimes, which is determined by the factor B^I_{hom} or expressed equivalently in terms of a critical packing density, depends on the solid friction coefficient and the modeled microstructure. The ability of the LCC model to capture such transition allows the consideration of the entire range of Drucker-Prager frictional behaviors. For the macroscopic scale, the model quantifies the effects of the amount of rigid inclusions and of different microstructures and interface conditions on the strength predictions.

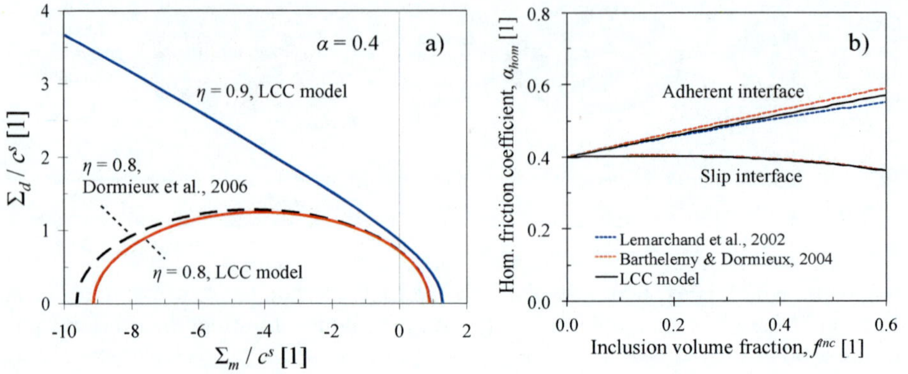

Fig. 2. a) Strength domains predicted for a porous solid with granular microstructure. The critical packing density marking the transition between elliptical and hyperbolic domains is $\eta_{cr}(\alpha{=}0.4) = 0.85$. b) Homogenized friction coefficients for a solid reinforced with rigid inclusions. The microstructure follows a matrix-inclusion morphology and perfectly adherent and slip interface conditions.

5 CONCLUSIONS

The LCC method represents a robust modeling framework for developing micromechanics solutions for the multiscale strength responses of cohesive-frictional materials. The results for benchmark microstructures compare adequately with existing micromechanics formulations, although further validation between micromechanics solutions and numerical methods is necessary to properly delimit the applicability of these methods for strength upscaling.

REFERENCES

[1] Suquet, P.: Effective behavior of nonlinear composites. In: Continuum Micromechanics, pp. 197–264. Springer, Heidelberg (1997)

[2] Lemarchand, E., Ulm, F.-J., Dormieux, L.: Effect of inclusions on friction coefficient of highly filled composite materials. J. Eng. Mech. 128(8), 876–884 (2002)

[3] Barthélémy, J.-F., Dormieux, L.: A micromechanical approach to the strength criterion of Drucker-Prager materials reinforced by rigid inclusions. Int. J. Num. Anal. Meth. Geomech. 28(7-8), 565–582 (2004)

[4] Dormieux, L., Kondo, D., Ulm, F.-J.: Microporomechanics. Wiley, Chichester (2006)

[5] Gathier, B., Ulm, F.-J.: Multiscale strength homogenization: application to shale nanoindentation, CEE Research Report R08-01. Massachusetts Institute of Technology, Cambridge (2008)

[6] Ortega, J.A., Gathier, B., Ulm, F.-J.: Homogenization of cohesive-frictional strength properties of porous composites: Linear comparison composite approach. J. Nanomech. Micromech. 1(1), 11–23 (2011)

[7] Ponte Castañeda, P.: Second-order homogenization estimates for nonlinear composites incorporating field fluctuations. J. Mech. Phys. Solids 50(4), 737–757, 759-782 (2002)

DIFFUSE BIFURCATIONS OF POROUS MEDIA UNDER PARTIALLY DRAINED CONDITIONS

WaiChing Sun[1,*] and José E. Andrade[2]

[1] Theoretical and Applied Mechanics
Northwestern University
A-161 Tech Institute, 2145 Sheridan Road
Evanston, IL, U.S.A.
*e-mail: steve.sun@u.northwestern.edu
[2] Division of Engineering and Applied Science
California Institute of Technology
MC 104-44, 221 Thomas Labs
Pasadena, CA, U.S.A.
e-mail: jandrade@caltech.edu

Summary. This paper presents a procedure which utilizes an energy functional of the apparent solid skeleton (solid skeleton + fluid/solid interface) and higher- order mixed-field finite element to assess the stability of porous media under partially drained conditions. Assuming that the porous medium is fully saturated and the solid skeleton deformation is infinitesimal, we show that the static equilibrium and the sufficient conditions for the onset of diffuse bifurcations can both be derived from the same energy functional in an incremental fashion. A three-dimensional example is included to demonstrate the predictive capacity of the proposed procedure.

Keywords: bifurcation, diffuse instability, mixed finite element method, poromechanics.

1 INTRODUCTION

It is well known that static liquefaction occurs in sand when the deviatoric stress reachs its peak value. While carefully prepared sand specimens under undrained triaxial compression tests can deform relatively homogeneous prior to reaching the peak value of deviatoric stress, the deformation of the specimens seldom remains homogeneous throughout [3]. Depending on the compactness of the sand, this inhomogeneous deformation may be followed by shear banding or a residual state in which spatially periodic deformed configuration may continuously develop without altering the imposed loading condition [8]. The goal of this study is to implement a procedure to predict and replicate these inhomogeneous deformation modes by modeling sand as a fully saturated porous medium.

2 VARIATIONAL FRAMEWORK

When a porous medium is subjected to loading under the partially drained condition, the solid skeleton deforms and the pore-fluid diffuses simultaneously. Due to the

R.I. Borja (Ed.): Multiscale and Multiphysics Processes in Geomechanics, SSGG, pp. 61–64.
springerlink.com

diffusion of pore-fluid, constitutive responses of the porous medium become gradient- and rate-dependent. While these features may regularize the governing equations for finite relaxation times, the non-local nature introduced by the pore-fluid diffusion process also makes it impossible to detect bifurcations on the material point level at a given time. Furthermore, since the motion of the solid skeleton and pore-fluid constituents are not identical under partially drained conditions, systems composed of porous media (solid skeleton + pore-fluid + fluid/solid interface) are open systems in which both matter and energy can exchange across the boundaries [5]. For those open systems, energy functional cannot be properly defined due to the mass exchange of pore-fluid.

To overcome the limitations alluded above, we consider an alternative close system composed of the apparent solid skeleton (solid skeleton + fluid/solid interface) where energy stored in pore-fluid due to compression is removed [6]. The upshot of the apparent solid skeleton system is that it allows one to define an energy functional whose Euler-Lagrange equation yields the static equilibrium equations. This valuable feature allows one to determine the onset of bifurcation modes by checking whether the extremum of the energy functional is attainable — if the extremum is attainable, the system is stable; otherwise, spurious zero-energy deformation modes would occur as bifurcation modes. Since porous media consolidate in a transient fashion, we discretize the energy functional in time via backward Euler scheme such that the total energy stored in the apparent solid skeleton can be expressed in an incremental form [7], i.e.,

$$\text{Total free energy } |_n = \text{Total free energy } |_{n+1} + \text{Dissipation}|_n^{n+1} \tag{1}$$

The dissipation in equation (1) stems from both constituents — the evolution of the solid skeleton internal variable and the diffusion of the pore-fluid. Meanwhile, the free energy at $t = t_{n+1}$ stems from the reversible work of the apparent solid skeleton. Once the expression of the total energy is known, it becomes trivial to obtain the discrete variational form of the static equilibrium equation from the first Gateaux variation and the second-order work from the second Gateaux variation of the total energy functional. By using the solid skeleton displacement $\boldsymbol{u}$ and pore pressure p as the generalized coordinates, the necessary condition for the stability of the porous system can be expressed as,

$$\mathcal{P}^s(\boldsymbol{u}, p + \delta p) \leq \mathcal{P}^s(\boldsymbol{u}, p) \leq \mathcal{P}^s(\boldsymbol{u} + \boldsymbol{\delta u}, p) \tag{2}$$

where $\mathcal{P}^s$ is the total free energy of the apparent solid skeleton. On the other hand, bifurcation modes are attained if one or both of the inequalities become equalities. In a Galerkin finite element setting, assuming that the trial and solution spaces are carefully chosen such that the Ladyshenskaya-Babuška-Brezzi (LBB) condition is satisfied [2, 4], diffuse bifurcation modes can be signaled by the singularity of the Hessian matrix of the second-order work of the apparent solid skeleton.

3 RESULTS AND DISCUSSION

To demonstrate how diffuse bifurcations can be captured in a finite element setting, a globally undrained triaxial compression simulation on sand is conducted. This finite element model is composed of 216 three-dimensional Taylor-Hood finite elements [4]. Each finite element contains 27 solid skeleton displacement and 8 pore pressure nodes such that the LBB condition is satisfied. In the finite element simulation, displacement is prescribed on the top and bottom boundaries. Normal traction is applied on the side boundaries. The boundaries are undrained but local drainage is allowed within the specimen. To replicate the constitutive response of the solid skeleton, a two-invariant pressure sensitive elasto-plasticity model called NOR-SAND is used [1]. Gravity is introduced as a small perturbation that breaks perfect homogeneity and the evolution of stress and strain in time is studied. Figure 1 depicts the evolution of deviatoric stress and axial strain. Prior to reaching the bifurcation point, deformation of the numerical specimen is approximately homogeneous despite of the presence of gravity. After the bifurcation point has been reached, it is clearly shown in Figure 1 that homogeneous deformation is no longer sustainable in the numerical specimen. Instead, the spatial pattern develops and follows by the formation of deformation bands. To identify the unstable spatial domain, we compute the condition number of the Hessian matrix of each finite element, i.e.,

$$\frac{1}{c} = \left| \frac{min(\lambda_i)}{max(\lambda_i)} \right| \; ; \; \lambda_i = \text{eigenvalues of the Hessian matrix} \tag{3}$$

Figure 2 depicts the the distributions of the pore pressure and the inverse of the condition number within the numerical specimen. It is shown that the zone with lowest $1/c$ is where deformation band forms. This result suggests that the unstable zone can be identified quantitatively even under partially drained condition. Such a feature is useful for multi-scale analysis where micro-mechanics enhancements are incorporated either concurrently or hierarchically in the unstable zone.

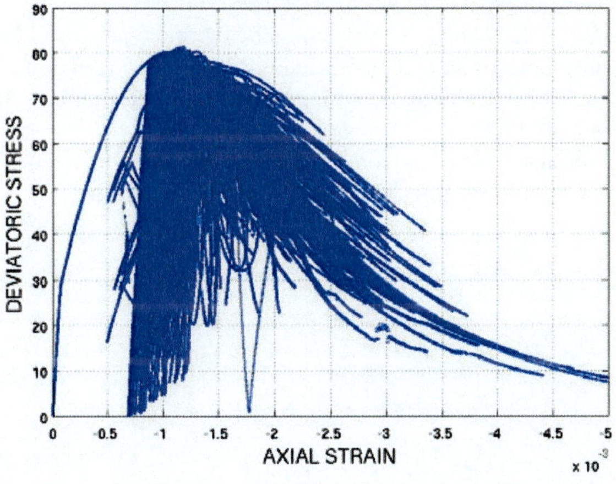

Fig. 1. Deviatoric stress vs. axial strain at all Gauss points in the finite element mesh.

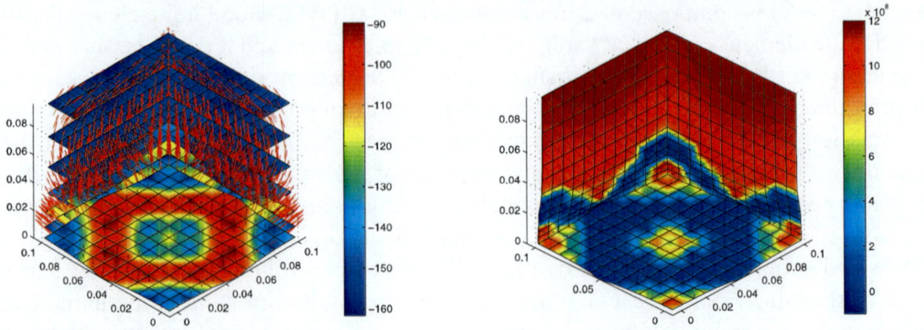

(a) Flow vectors at the end of simulation (b) Deformed configuration at the end of simulation

Fig. 2. Finite element simulation of the undrained triaxial compression test. The color contours represent the (a) pore pressure distribution and (b) the inverse of the condition number $1/c$ of the Hessian matrix.

REFERENCES

[1] Andrade, J.E., Borja, R.I.: Modeling deformation banding in dense and loose fluid-saturated sands. Finite Elements in Analysis and Design 43, 361–383 (2007)
[2] Bardet, J.P.: Finite element analysis of two-phase porous hypoelastic solids under plane strain loadings. Engineering Computations 13, 29–48 (1996)
[3] Borja, R.I.: Condition for liquefaction instability in fluid-saturated granular soils. Acta Geotechnica 1, 211–244 (2006)
[4] Bathe, K.J.: The inf-sup condition and its evaluation for mixed finite element methods. Computers and Structures 79, 243–252 (2001)
[5] Coussy, O.: Poromechanics. Wiley, New York (2004)
[6] Coussy, O., Pereira, J.-M., Vaunat, J.: Revisiting the thermodynamics of hardening plasticity for unsaturated soils. Computers and Geotechnics 37, 207–215 (2010)
[7] Simo, J.C., hughes, T.J.R.: Computational Inelasticity, Interdisciplinary Applied Mathematics. Springer, New York (2000)
[8] Vardoulakis, I.: Shear-banding and liquefaction in granular materials on the basis of a Cosserat continuum theory. Ingenieur-Archiv 59, 106–113 (1989)

DISCONTINUOUS PORE FLUID DISTRIBUTION UNDER MICROGRAVITY DUE TO PARTICLE REARRANGEMENT

Ming Xiao[1,*], Lakshmi N. Reddi[2], and Susan L. Steinberg[3]

[1] Department of Civil and Geomatics Engineering, M/S EE94. California State University, Fresno, CA 93740. USA
*e-mail: mxiao@csufresno.edu, web page: http://www.csufresno.edu
[2] Department of Civil and Environmental Engineering, University of Central Florida, Orlando, FL 32816. USA
[3] Universities Space Research Association, Mail Code EC3, Building 29, Room 110, NASA Johnson Space Center, 2101 NASA Rd 1, Houston, TX 77058, USA

Summary. This paper reports (i) the experimental investigation conducted aboard the NASA's KC-135 reduced-gravity flights to study possible particle separation and the distribution of discontinuous wetting fluid in porous media under microgravity, and (ii) numerical study of the effect of soil particle rearrangement on the water retention characteristics in zero gravity.

Keywords: microgravity, network model, particle rearrangement, pore fluid, porous media, water retention characteristics.

1 INTRODUCTION

The study reported in this paper is part of a long-term fundamental research investigation involving design of plant growth systems for crop production under microgravity. Spatial distribution of wetting fluid as well as size distribution of discontinuous pore fluid ganglia under microgravity may strongly impact plant growth conditions, such as water transport and intake by the root in the soil substrate. With gravity, excess water drains along a vertical gradient, and water recovery is easily accomplished; under microgravity, the distribution of water is less predictable and can easily lead to localized over-moistening or anoxia [1].

2 EXPERIMENTAL INVESTIGATION

2.1 Methodology

Parabolic flight of KC-135 aircraft offers researchers reduced gravity in the order of 10^{-2} g for up to 25 sec. The short microgravity period facilitates investigations on the role of gravity in fundamental physical processes. In this study, glass beads were used as porous media. Hexadecane, a petroleum compound immiscible with and lighter than water, was used as wetting fluid at residual saturation in glass beads. The higher freezing point of hexadecane (18°C) allowed the relatively quick onboard solidification of discontinuous pore fluid entrapped in glass beads, so that their size distribution could be measured using wet sieving after the flight. The experiments were conducted onboard NASA's KC-135 flight in February 2004. Figure 1 depicts a representative flying profile of the KC-135 flight. The plane simulated varying gravity conditions by flying in parabolic profile, including a 25-sec 1.8g environment while gaining altitude at a

R.I. Borja (Ed.): Multiscale and Multiphysics Processes in Geomechanics, SSGG, pp. 65–68.
springerlink.com

45-degree angle upward, a 25-sec microgravity environment at the top of the parabola, and another 25-sec1.8g condition when the plane dives downwards. A nitrogen freezer (–150°C) was used to freeze the hexadecane pore fluids under 1g, 1.8g, and 10^{-2} g. Within the wall matrix of the freezer, liquid nitrogen was adsorbed and could spill out under microgravity. Three samples were put into the freezer prior to the 25s of microgravity — one before flight's takeoff, one during the cruising of the flight (1g), and one when 1.8g was reached in the cabin. During the first 10 sec of the 25-sec microgravity, the glasses beads that encapsulated hexadecane pore fluid were free to rearrange, and then the capsule that contained the specimen was carefully put in the freezer that was able to freeze the hexadecane liquid in 15 sec and therefore preserve the changes that occurred in microgravity. Four KC-135 flights were conducted with one flight per day. The same procedure was repeated during the other 3 days of flights to test the reproducibility. After the flights, the measurements of the blob size distribution (BSD), which simulated the pore size distribution corresponding to residual saturation, were conducted in a constant-temperature room at 6°C. Then statistical analyses using the Kolmogorov-Smirnov test were conducted to verify whether the BSDs under the gravity conditions were significantly different from each other.

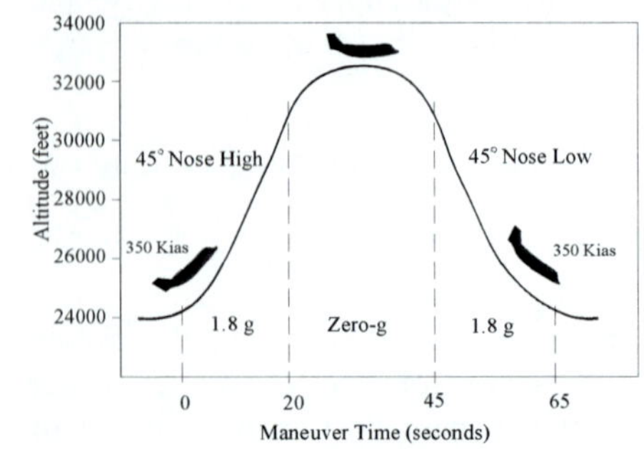

Fig. 1. The trajectory of the KC135 aircraft

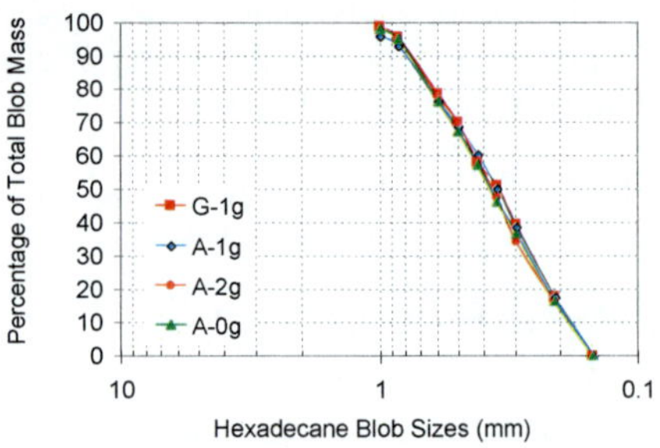

Fig. 2. Cumulative blob size distributions

2.2 Experimental Results and Analyses

The measured hexadecane BSDs of the 16 samples are presented in terms of cumulative distributions. Figure 2 shows representative four BSDs from the Day-4 experiments, in which the BSDs are the pore fluid distributions under 1g on ground (labeled as G-1g), 1g in the air (A-1g), 1.8g in the air (A-2g) and microgravity in the air (A-0g). The Kolmogorov-Smirnov analysis concluded that the pore fluid distributions did not change when the gravity condition changed from 1g on ground to 1g in the air, from 1g to 1.8g, and from 1.8g to 10^{-2}g. Discontinuous pore fluid ganglia showed no noticeable breakup or coalescence during microgravity.

3 THEORETICAL INVESTIGATION

3.1 Model Formulation

In order to provide an understanding of the effect of particle rearrangement on water retention characteristics under microgravity, a three-dimensional pore network model is developed, based on a two-dimensional network model proposed by Payatakes et al. [2], Ng et al. [3], and Payatakes et al. [4]. In a porous medium, an irregular pore can be idealized as a pore body with pore throats (Figure 3a). The pore body diameter, the pore length, and the volume of the unit cell can be calculated using the pore throat diameter. The formulae were presented by Payatakes et al. [4]. The pore network is made of three-dimensional elemental pores, i.e., conceptual elemental void space (CEVS). Each CEVS consists of six half-unit-cells in the three dimensions. The six half-unit-cells in a CEVS can have different sizes, and together they make up of one 3D conceptual pore. Figure 3b shows a 3D pore network with dimension of 2×2×2 CEVSs. This study uses Ottawa sand whose particles are spheroid and have rough surface. The pore size distributions are represented using the pore throat distributions, which are derived from grain size distribution based on two extreme particle arrangements—the most compact rhombohedral and the most porous cubic packing. The model is then used to reveal the variation of water retention characteristics and the spatial distribution of pore fluid due to particle rearrangements.

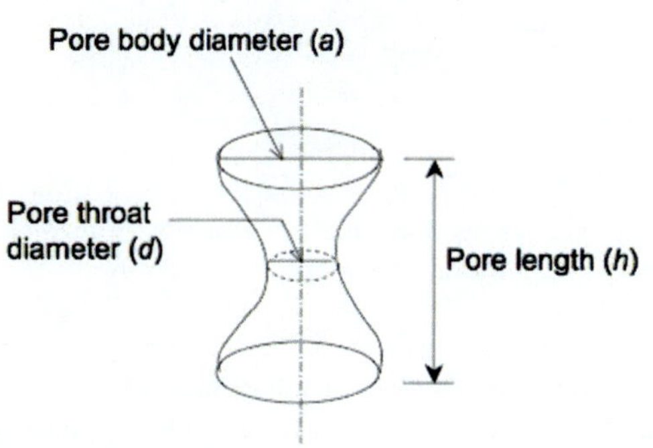

Fig. 3a. Unit cell simulating a pore

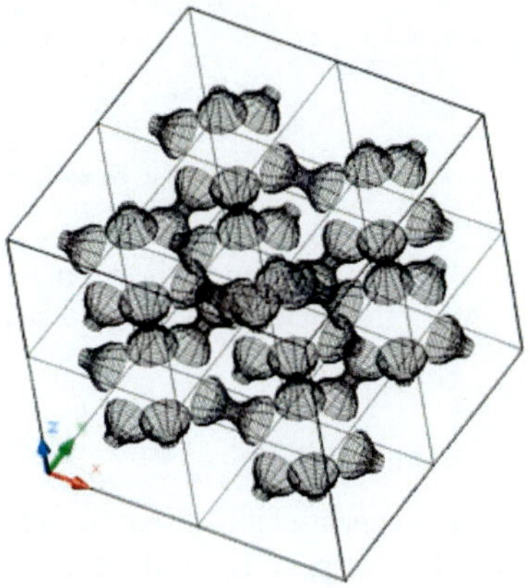

Fig. 3b. 3D pore network model (2×2×2 CEVS)

3.2 Model Results

Particle rearrangement under zero gravity could happen *during* the drying or rewetting process, resulting in a "dynamic" pore size distribution and temporal change of water retention characteristics. This model studies the changes in water retention characteristics of Ottawa sand during the drying process due to particle rearrangement. The Ottawa sand is initially saturated under the cubic packing. At suction head of 10cm and volumetric water content of 0.46, the particle arrangement changes to rhombohedral packing. After that, the drying process continues under the new rhombohedral packing. Figure 4 shows that after the particle rearrangement the drying curve follows the drying curve of the original rhombohedral-packing. If the particle rearrangement (from cubic to rhombohedral packing) occurs at suction head of 14cm when the volumetric water content was close to the residual water content, the drying curve changes little and it continues to follow the water retention curve under the cubic packing. The simulations indicate that water retention characteristics are less affected by the particle rearrangement at low water content than at high water content. Figure 5 shows the predicted pore water spatial distributions in 3D pores (10×10×5 CEVS) during a drying process under 0g. Figure 5a shows the spatial distribution of pore water (represented in black) in the pore network in the cubic packing, in which 36 CEVSs are occupied by water; Figure 5b shows the pore water distribution after the particles re-arrange from cubic to rhombohedral packing at a suction of 14 cm during the drying process and 52 CEVSs are occupied by water. Figure 5 suggests that the pore fluid is not uniformly distributed. The model results are in agreement with the study by Podolsky and Mashinsky [5] who observed local over-moistening in a substrate sample aboard the "Salyut-7" and "Mir" space stations.

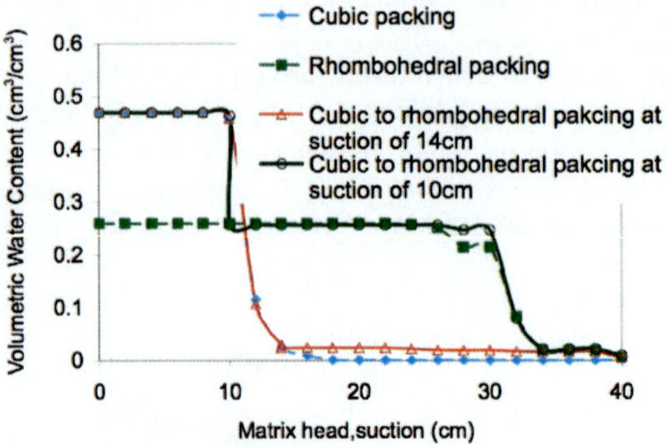

Fig. 4. Predicted changes in water retention of Ottawa sand due to particle rearrangement during drying process, under 0g.

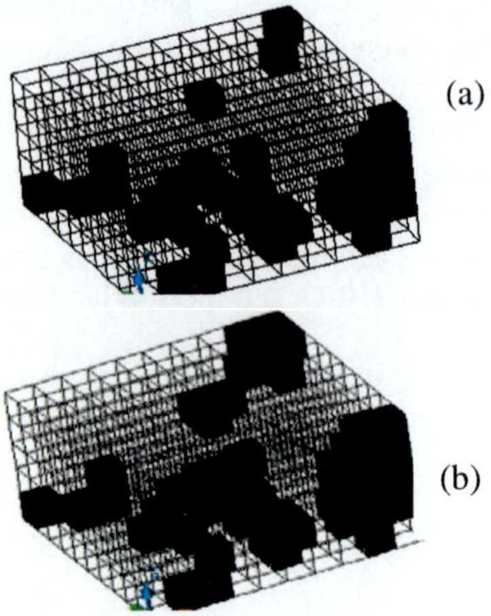

Fig. 5. Predicted spatial distributions of pore fluid due to particle rearrangement during drying process, under 0g.

4 CONCLUSIONS

The experimental research indicates microgravity has little effect on the size distribution of pore fluid blobs corresponding to residual saturation of wetting fluids in porous media. The blobs showed no noticeable breakup or coalescence during microgravity. The model simulations indicate increased hysteresis when particle arrangement changes from loose condition to compact condition. Particle rearrangement (from cubic to rhombohedral packing) significantly alters the water retention characteristics when the porous medium is nearly saturated. When the porous medium is close to residual saturation, water retention characteristics are almost unaffected by the particle rearrangement. Pore fluid spatial distribution in the porous media (Ottawa sands) is found to be non-uniform under zero gravity.

REFERENCES

[1] Hoehn, A., Scovazzo, P., Stodieck, L.S., Clawson, J., Kalinowski, W., Rakow, A., Simmons, D., Heyenga, A.G., Kliss, M.H.: Microgravity root zone hydration systems. In: 30th International Conference on Environmental Systems, Toulouse, France. SAE Technical Paper Series, 2000-01-2510 (2000)

[2] Payatakes, A.C., Tien, C., Turian, R.M.: A new model for granular porous media: part I. Model formulation. AIChe J. 19(1), 58–67 (1973)

[3] Ng, K.M., Davis, H.T., Scriven, L.E.: Visualization of blob mechanics in flow through porous media. Chem. Engrg. Sci. 33, 1009–1017 (1978)

[4] Payatakes, A.C., Ng, K.M., Flumerfelt, R.W.: Oil ganglion dynamics during immiscible displacement: Model formulation. AIChe J. 26(3), 430–442 (1980)

[5] Podolsky, I., Mashinsky, A.: Peculiarities of moisture transfer in capillary-porous soil substitutes during space flight. Adv. Space Research 14(1), (11)39-(11)46 (1994)

SCALE EFFECT OF CAVITY EXPANSION IN SOIL

Samira Ladjal and Wei Wu*

Institut für Geotechnik, Universität für Bodenkultur, 1180 Vienna, Austria
* e-mail: wei.wu@boku.ac.at, web page: http://www.baunat.boku.ac.at/489.html

Summary. In this paper the growth of plant roots under increasing mechanical impedance is investigated. A mechanical model based on the expansion of cavity is used for the penetration of plant roots into compacted soils. The soil is modelled by an enhanced plastic constitutive equation with strain gradient. Our model can be used to explain some scale effect in the interaction between plant root and soil.

Keywords: cavity expansion, scale effect, strain gradient.

1 INTRODUCTION

An important factor controlling the growth and yield of plants is the ability of the roots to penetrate through soil to acquire water and nutrients, which in turn has substantial influence on the development of shoots and yields. The relation between root growth and soil impedance has been a subject of intensive research ([2], [4]). The growth of plant roots is characterized by axial penetration accompanied by radial expansion in the elongation zone. Root penetration was studied as cavity expansion by Farrel and Greacen [5]. Previous investigations on the penetration resistance of roots, e.g. Misra et al. [10, 11], were mainly based on the theory of cavity expansion by Vesic [15]. The soil surrounding the root was modelled by an elastic, perfectly plastic material. Recently, root penetration was investigated by Kirby and Bengough [7] using Critical State Model and FEM.

In soil science, the penetration tests are used to characterize the compaction of agriculture soils. The penetrometers have typically a diameter between 10 cm and 20 cm, which are much larger than the usual root diameter, between 0.2 mm and 2 mm. Some laboratory tests with small penetrometers show clear dependence of the penetration pressure depends on the penetrometer diameter ([16]). This scale dependence is elusive to the classical theory of cavity expansion, since the constitutive equation does not involve any length scale.

2 EXPERIMENTAL EVIDENCE

The experimental results by Whiteley and Dexter [16] and by Bolton et al. [3] are evaluated and shown in Figure 1. The tests by Bolten and his co-workers were carried out in a geotechnical centrifuge. The soil used by Whiteley and Dexter is a silty soil with a mean grain diameter of about 0.1 mm. The specimen was prepared by remolding the soil at the water content close to its plastic limit of about 17.3%. The soil used by Bolton and his co-workers is clean sand with a mean grain diameter of

R.I. Borja (Ed.): Multiscale and Multiphysics Processes in Geomechanics, SSGG, pp. 69–72.
springerlink.com

about 0.9 mm. The penetration resistance in Figure 1a is obtained by normalizing the penetration force with the cross-section area of the penetrometers. The normalized resistance in Figure 1b is defined by ratio between the penetration pressure and the overburden stress.

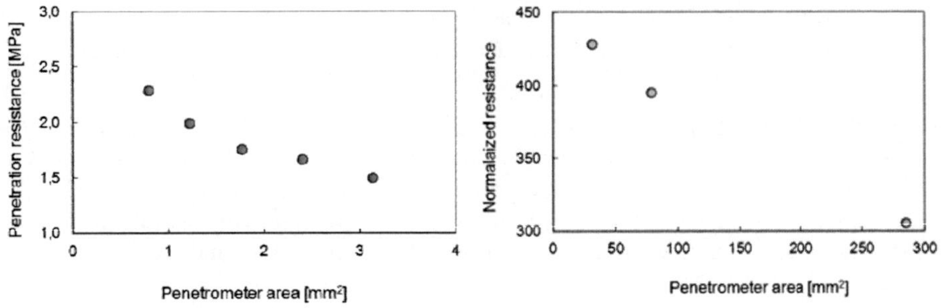

(a) (b)

Fig. 1. Penetration tests (a) by Whiteley and Dexter [16] and (b) by Bolton et al. [3]

In both tests the penetration resistance decreases with increasing penetrometer diameter. For large diameter the penetration resistance is expected to approach a limit value, which can be obtained from the classical theory of cavity expansion. Obviously this scale dependence cannot be predicted by the classical theory, since the constitutive equation does not contain any term with length scale.

3 STRAIN GRADIENT MODEL

In order to account for the scale effect, enhanced constitutive equations are needed by including higher order terms. The higher order terms give rise to the length scale, which can be correlated to some material properties like the mean grain diameter of soil. There are mainly three approaches to introduce higher order terms into the constitutive equation, i.e. the Cosserat theory ([11]), nonlocal theory ([1]) and strain gradient theory ([14]). We will make use of the strain gradient theory by Vardoulakis and Aifantis ([14]). The theoretical framework is briefly recapitulated below. Consider the Drucker-Prager yield function

$$f(\sigma, k) = \alpha I_1 + \sqrt{J_2} - k = 0 \tag{1}$$

where $I_1 = \sigma_{ii}$ is the first invariant of stress tensor and $J_2 = s_{ij}s_{ij}/2$ is the second invariant of the stress deviator. The hardening parameter k is assumed to depend on the higher order of strain gradient

$$k = k(\varepsilon_e) + c\nabla^2\varepsilon_e \tag{2}$$

where $\varepsilon_e = \sqrt{2\varepsilon_{ij}\varepsilon_{ij}/3}$ is the effective plastic strain and c is a material parameter. The gradient term in (2) requires an additional boundary condition $(\nabla\varepsilon_e)_i n_i = 0$ with n_i

being the outward normal to the boundary ([12]). The above formulation represents the simplest version of gradient theory by introducing the higher order term into the yield function while leaving the other features of plasticity theory unchanged.

4 EXPANSION OF CYLINDRICAL CAVITY

Consider a cylinder with thick wall subjected to a uniform internal pressure p. The cylinder is assumed to be infinitely long so that the plane strain condition along its axis is valid. Since both the geometry and the load are axisymmetric, all field variables depend only on the radial displacement u. The axial and the circumferential displacements are zero. Moreover, there are only two non-vanishing strain components, i.e. the radial strain ε_r and the circumferential strain ε_θ. The problem formulation includes the equilibrium equation, the constitutive equation and the boundary conditions. The problem can be further simplified by assuming that the material is not compressible. This is a reasonable assumption for most agriculture soils. In this case the radial displacement can be written out $u = C / r$ with C being an integration constant. The resulting differential equation can be solved with the relaxation method. The numerical results show that the limit pressure depends on the cavity diameter a. For large cavity diameter, the solution approaches the classical solution asymptotically. By relating the material parameter c in (2) with the mean grain diameter of soil, the scale effect in the experiments can be well reproduced.

5 IMPLICATION FOR ROOT GROWTH

The above theoretical model has some interesting implications for the growth of root in soil with high impedance. The seminal roots of most plants thicken when growing into soil with higher mechanical impedance. According to the classical theory of cavity expansion, an increase in root diameter does not bring any benefit since the penetration force is proportional to the cross section of the root. It would be more economic for the root to increase the penetration force rather than the diameter. There are two conjectures as to why roots thicken under increasing impedance. Farrel and Greacen [5] suggested that the larger radial expansion of thicker root may cause cracking near the root tip, which in turn makes root penetration easier. Logsdon et al. [8] showed that thicker roots possess higher strength against buckling. However, none of these conjectures is verified by experiments.

Our study provides a further explanation to root thickening. Larger root diameter gives rise to lower penetration pressure. The driving force is known to be provided by the turgor pressure in the root cells. When faced with increasing mechanical impedance, plant roots may choose either to increase the turgor pressure while keeping the root diameter unchanged or to increase the root diameter while keeping the turgor pressure unchanged. Plant root may of course increase both the diameter and the turgor pressure simultaneously. Note that the turgor pressure (about 1 MPa) is not unlimited and is mainly dictated by the behaviour of the cell walls. Therefore, it makes sense to increase the root diameter instead of the turgor pressure. Some experiments in plant science indicate that the turgor pressure does not show remarkable change when roots are mechanically impeded. For constant turgor pressure, the penetration force will be proportional to the cross section area of the roots. However, root growth is enhanced by the reduced penetration resistance from the soil. Finally, it should be noticed that the interaction between root and soil is not purely mechanical. In general, compaction changes not only the mechanical strength of soil but also its water potential, which requires hydro-mechanically coupled analyses.

REFERENCES

[1] Bazant, Z.P., Pijaudier-Cabot, G.: Nonlocal continuum damage, localization instability and convergence. J. Applied Mechanics, Transactions ASME 55, 287–293 (1988)

[2] Bengough, A.G., Mullins, C.E.: Mechanical impedance to root growth: a review of experimental techniques and root growth responses. J. Soil Science 41, 341–358 (1990)

[3] Bolton, M.D., Gui, M.W., Garnier, J., Corte, J.F., Bagge, G., Laue, J., Renzi, R.: Centrifuge cone penetration tests in sand. Géotechnique 49, 543–552 (1999)

[4] Clark, L.J., Whalley, W.R., Barraclough, P.B.: How do roots penetrate strong soil? Plant and Soil 255, 93–104 (2003)

[5] Farrel, D.A., Greacen, E.L.: Resistance to penetration of fine probes in compressible soil. Australian J. Soil Research 4, 1–17 (1966)

[6] Hanbury, C.D., Atwell, B.J.: Growth dynamics of mechanically impeded lupin roots: Does altered morphology induce hypoxia? Annals of Botany 96, 913–924 (2005)

[7] Kirby, J.M., Bengough, A.G.: Influence of soil strength on root growth: Experiments and analysis using a critical-state model. European J. Soil Science 53, 119–127 (2002)

[8] Logsdon, S.D., Parker, J.C., Reneau Jr., R.B.: Root growth as influenced by aggregate size. Plant and Soil 99, 267–275 (1987)

[9] Misra, R.K., Dexter, A.R., Alston, A.M.: Penetration of soil aggregates of finite size - I. Blunt penetrometer probes. Plant and Soil 94, 43–58 (1986)

[10] Misra, R.K., Dexter, A.R., Alston, A.M.: Penetration of soil aggregates of finite size - II. Plant roots. Plant and Soil 94, 59–85 (1986)

[11] Mühlhaus, H.-B., Vardoulakis, I.: Thickness of shear bands in granular materials. Géotechnique 37, 271–283 (1987)

[12] Mühlhaus, H.-B., Alfantis, E.C.: A variational principle for gradient plasticity. International J. Solids and Structures 28, 845–857 (1991)

[13] Richard, B.G., Greacen, E.L.: Mechanical stresses on an expanding cylindrical root analogue in granular media. Australian J. Soil Research 24, 393–404 (1986)

[14] Vardoulakis, I., Aifantis, E.C.: A gradient flow theory of plasticity for granular materials. Acta Mechanica 87, 197–217 (1991)

[15] Vesic, A.S.: Expansion of cavities in an infinite soil mass. J. Soil Mechanics and Foundation Division, ASCE 98, 265–290 (1972)

[16] Whiteley, G.M., Dexter, A.R.: The dependence of soil penetrometer pressure on penetrometer size. J. Agricultural Engineering Research 26, 467–476 (1981)

THE ROLE OF MODELING AND SIMULATIONS IN ESTIMATING MULTISCALE EFFECTIVE PERMEABILITY

P.A. Selvadurai* and A.P.S. Selvadurai**

Department of Civil Engineering and Applied Mechanics
McGill University
817 Sherbrooke Street West, Montréal, QC, Canada H3A 2K6
*e-mail: pa.selvadurai@berkeley.edu
**e-mail: patrick.selvadurai@mcgill.ca
web page: http://www.mcgill.ca/civil/faculty/selvadurai/

Summary. The paper describes the procedures developed to measure the effective permeability of a large cuboidal region of Indiana Limestone measuring 508 mm. A patch permeability test was performed to measure the permeability characteristics of a local region of the cuboid. The surface measurements are then used to estimate, via a kriging technique, the distribution of interior permeability within the block. The results of these investigations provide estimates for the representative permeabilities of 64 sub-cube regions of the cuboidal block. These are used in conjunction with computational modeling to assess the predictive qualities of various procedures for estimating the effective permeability of the cuboidal region. This paper proposes the applicability of the *geometric mean* as an estimate for the "*effective permeability*" of the hydraulically nearly isotropic Indiana Limestone.

Keywords: effective permeability, Indiana limestone, patch permeability tests, permeability.

1 INTRODUCTION

Permeability is a key parameter that occupies a central position in geomechanics and in the environmental geosciences. Solutions to problems dealing with groundwater hydrology, contaminant transport in the geosphere, deep geologic disposal of contaminants and other hazardous materials and mechanics of fluid-saturated deformable porous media are greatly influenced by the accuracy with which the permeability of a geomaterial can be estimated. In its classical continuum definition, permeability is a measure that is applicable to a mathematical point that occupies *no space*. In reality, however, permeability of a naturally occurring geomaterial is highly influenced by scale and while there are no universally accepted guidelines that can be used to identify the relevant scales; permeability can be influenced by regional scales of the order of 0.5 km to 5.0 km, to borehole scales ranging from 30 m to 300 m to laboratory scales ranging from 5 cm to 15 cm [1]. In naturally occurring geomaterials, the scale also introduces other anomalies including fractures, fissures, damage zones, voids and vugs that can continue to evolve depending on coupled geomechanical, geochemical and geomorphological processes.

Conventional laboratory procedures involving steady-state and/or transient tests conducted on recovered cores are perhaps the most widely used techniques for estimating permeability of geomaterials [2]. The use of large-diameter cores (450 mm) for permeability testing using either axial flow or radial flow tests can alleviate some of the

R.I. Borja (Ed.): Multiscale and Multiphysics Processes in Geomechanics, SSGG, pp. 73–76.
springerlink.com

drawbacks associated with scale in the permeability testing at the laboratory scale. These are however non-routine research class experiments [1]. When opportunities exist for testing larger cuboidal samples, techniques need be developed to determine the effective permeability characteristics of such samples. This research deals with the development of experimental procedures, mathematical analysis and computational modeling as a basis for developing an approach to estimate the effective permeability of a large sample of Indiana Limestone. When large samples can be acquired, the permeability testing has to be approached in such a way that the maximum information regarding the permeability can be obtained in a non-destructive manner. This challenge was presented in the testing of a large cuboid of Indiana Limestone measuring approximately 508 mm. The testing procedure involved the development of a surface permeability device, its mathematical and computational modeling, the evaluation of the distribution of surface permeability, the estimation of interior permeability of the cuboidal region and the use of the experimental data set to estimate the multiscale effective permeability of the cuboidal region.

2 EXPERIMENTAL TECHNIQUES

The experimental technique involved the application of a steady flow rate to a central cavity of an annular sealing region that was maintained in contact with the plane face of the cuboidal sample of Indiana Limestone. The annular sealing region permits the development of an effective seal between a permeameter and the plane surface of the Indiana Limestone with the minimum of external load and enables the modeling of the permeameter using both mathematical and computational approaches. The general arrangement of the experimental facility, which includes the reservoir containing the cuboidal block of Indiana Limestone, the permeameter, the reaction frame and the hydraulic system used to apply the sealing loads, a precision flow pump that supplies fluid to the permeameter to develop steady-state flow in rock and ancillary devices for measuring the fluid pressures and sealing loads, is shown in Figure 1. Further details of the test facilities and test procedures are described in [3].

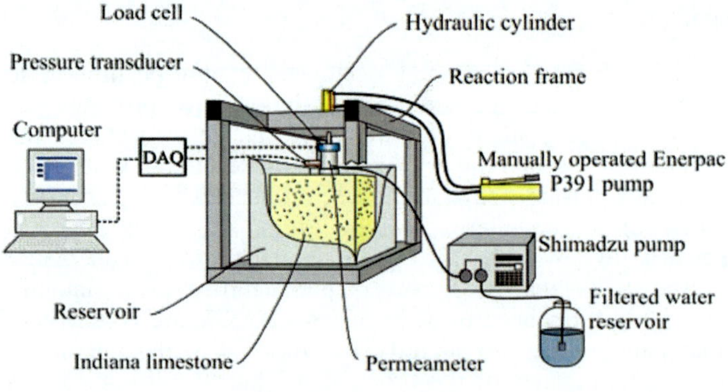

Fig. 1. The test facilities

3 ANALYTICAL AND COMPUTATIONAL MODELING

The mathematical modeling assumes the validity of Darcy flow and incompressible low Reynolds number flow in the connected pore space of the Indiana Limestone block. Furthermore, we assume that within the zone of influence of the permeameter, which is roughly equal to a hemi-spherical region corresponding to the outer diameter of the annular sealing zone, the effective permeability can be represented by an equivalent homogeneous value. The reduced Bernoulli potential $\boldsymbol{\Phi}(\mathbf{x})$ is thus governed by Laplace's equation

$$\nabla^2\Phi(\mathbf{x}) = 0 \tag{1}$$

where ∇^2 is Laplace's operator. The steady flow problem for the permeameter can be examined by solving (1). When the permeameter is located remote from the edges of the cuboidal region, the resulting three-part mixed boundary value problem can be reduced to the solution of a set of triple integral equations, which can be solved in an approximate manner [4]. In general, and when the permeameter is located in close proximity to the boundary of the cuboidal region, computational methods need to be used to relate the steady flow rate through the central cavity to the corresponding potential. The work reported in [4] indicates that the solution to the problem of an annular region located at the surface of a halfspace region can be used to estimate the local permeability of the tested location and that the steady flow rate Q is related to the differential potential Φ_0 applied to the central cavity of the annular region by the approximate relationship

$$Q = 4a\Phi_0\left(\frac{K\gamma_w}{\mu}\right)\left\{\begin{array}{l} 1+\left(\dfrac{4}{\pi^2}\right)c+\left(\dfrac{16}{\pi^4}\right)c^2+\left(\dfrac{64}{\pi^6}+\dfrac{8}{9\pi^2}\right)c^3+\left(\dfrac{64}{9\pi^4}+\dfrac{256}{\pi^8}\right)c^4 \\ +\left(\dfrac{92}{225\pi^2}+\dfrac{384}{9\pi^6}+\dfrac{1024}{\pi^{10}}\right)c^5+\mathrm{O}(c^6) \end{array}\right\} \tag{2}$$

where a is the inner radius of the annular seal, γ_w is the unit weight of water, μ is the dynamic viscosity, K is the intrinsic permeability, $c = a/b$ and b is the external radius of the annular sealing region. As $b \to \infty$, the result converges to the exact solution [5]. Computational modeling was conducted using a multiphysics finite element code and the accuracy of the computational treatments was established using the analytical result (2) for purposes of estimating the local permeability of the test location.

4 EXPERIMENTAL RESULTS AND EFFECTIVE PERMEABILITY

The permeabilities at 54 surface locations of the cuboidal region were first determined using the annular permeameter and the associated theoretical and computational developments. This data set was used in conjunction with a kriging procedure to develop estimates for the interior permeability distribution within the block of Indiana Limestone. Figure 2 illustrates the distribution of permeability within the cuboidal region. The permeabilities were then estimated for 64 sub-cube regions measuring 127 mm in dimension. The permeability in these sub-cube regions varied from a maximum of 250 $\times 10^{-15}$ m^2 to a minimum of 11×10^{-15} m^2. Also, it is observed that the spatial distribution of hydraulic properties conforms to a lognormal form. The permeability distribution associated with these sub-cube regions were used in conjunction with theoretical estimates for the '*effective permeability*' proposed in the literature. These included the Wiener bounds, the estimate developed using the results of Landau and Lifschitz and specific estimates for the effective permeability developed by Matheron, King and Dagan. These theoretical estimates for the effective permeability of the cuboidal region give a maximum effective permeability of 80 $\times10^{-15}$ m^2 and a minimum effective permeability of 71 $\times10^{-15}$ m^2 and the details of the analysis are given in [4]. To assess the accuracy of these theoretical estimates, computational estimates of one-dimensional permeability were determined by

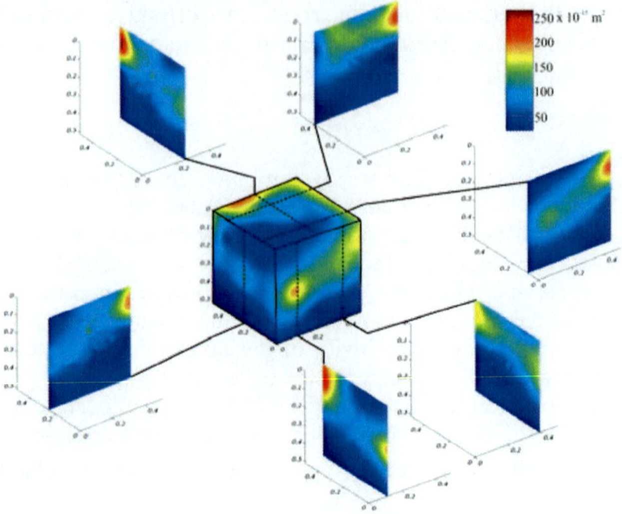

Fig. 2. The spatial distribution of the permeability in the cuboidal sample determined using the surface values of experimental data and kriging procedures.

subjecting the assemblage of sub-cube regions to one-dimensional flow, successively in three orthogonal directions. The values for the permeabilities were $K_1 = (78.44) \times 10^{-15}\ \mathrm{m}^2$, $K_2 = (78.98) \times 10^{-15}\ \mathrm{m}^2$ to $K_3 = (64.76) \times 10^{-15}\ \mathrm{m}^2$, indicating the presence of an extremely weak form of transverse isotropy, that permits the approximation of the effective permeability of the cuboidal region of Indiana Limestone by the geometric mean, $K_{eff}^S = \sqrt[3]{K_1 K_2 K_3} \simeq 74 \times 10^{-15}\,\mathrm{m}^2$.

5 CONCLUSIONS

The paper describes a basic methodology that can be adopted to estimate the effective permeability characteristics of large samples of porous geomaterials. The procedure involved the development of suitable experimental methods for conducting localized flow tests, their analysis and the extrapolation of the surface measurements to estimate the interior distributions of permeability. The approach is demonstrated through experiments conducted on a large cuboidal block of Indiana Limestone. The research methodology can be extended to cover other irregular regions of geomaterials that are amenable to application of either steady flow or transient hydraulic pulse tests.

REFERENCES

[1] Selvadurai, A.P.S., Boulon, M.J., Nguyen, T.S.: The permeability of an intact granite. Pure and Appl. Geophys. 162, 373–407 (2005)

[2] Selvadurai, A.P.S.: The influence of residual hydraulic gradients on decay curves for one-dimensional pulse tests. Geophysical J. Int. 177, 1357–1365 (2009)

[3] Selvadurai, P.A.: Permeability of Indiana Limestone: Experiments and Theoretical Concepts for Interpretation of Results, M. Eng. Thesis, McGill University (2010)

[4] Selvadurai, A.P.S., Selvadurai, P.A.: Surface permeability tests: Experiments and modelling for estimating effective permeability. In: Proc. Roy. Soc. A. Math. Phys. (2010), doi: 10.1098/ rspa.2009.0475

[5] Selvadurai, A.P.S.: The analytical method in geomechanics. Appl. Mech. Rev. 60, 87–106

BIOGROUT PROPAGATION IN SOILS

L. Laloui* and S. Fauriel

Soil Mechanics Laboratory
Swiss Federal Institute of Technology
EPFL-ENAC-LMS
Station 18
CH-1015 Lausanne
* e-mail: lyesse.laloui@epfl.ch, suzanne.chalindar@epfl.ch, web page: http://lms.epfl.ch/en

Summary. Microbial Induced Calcite Precipitation (MICP) is an innovative technique for soil grouting in which injected bacteria supplemented with chemicals lead to calcite precipitation in soils. A comprehensive research work is carried out to better understand and describe the phenomenon of miscible and reactive biogrout injection in saturated, deformable soil. A mathematical model accounting for the bio-hydro-mechanical couplings existing between fluid flow, bacterial concentration evolution, exchanges between phases, transport and miscibility is developed.Finite element modelling of column injection tests is carried out to validate the formulation and to demonstrate the potentiality of the developed model.

Keywords: bio-hydro-mechanical modelling, finite element, grouting, transport of bacteria.

1 INTRODUCTION

The growing need for infrastructure creates challenges in identifying new locations and creating more innovative construction techniques. Usually, chemical and cement based grouting products,leading to permanent soil pollution or emission of carbon dioxide,are used for soil improvement. More environmentally friendly grouting products such as biogrouting are therefore in demand. There is, however, a need to develop a comprehensive bio-hydro-mechanical model able to realistically predict the behaviour of biogrouted soils.

2 MODEL DEVELOPMENT

Based on bacterial and grouting fluid transport models [1-6], the following processes are considered in the developed model:(i) bacterial transport (advection, diffusion and dispersion and, additionally, random motility and chemotaxis), (ii) bacterial growth/decay,(iii) attachment/detachment,(iv) miscibility and (v) hydro-mechanical coupling.

The development is carried out in the framework of continuous porous media based on the notion of a Representative Elementary Volume (REV) that allows establishing balance equations at the macroscale. Figure 1 represents a REV before and after biogrout injection. Assuming that the porous medium is saturated, an injected REV is composed of two phases, the solid phase s consisting of the solid grains and the attached bacteria and the fluid phase f corresponding to the pore water and the suspended bacteria.

R.I. Borja (Ed.): Multiscale and Multiphysics Processes in Geomechanics, SSGG, pp. 77–80.
springerlink.com

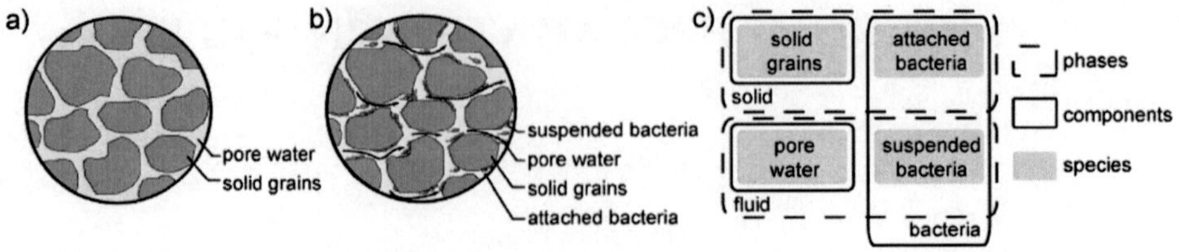

Fig. 1. (a) REV without bacteria, (b) REV with bacteria and (c) phases, components and species

3 BIO-HYDRO-MECHANICAL MODEL

3.1 Balance Equations

In the development of a bio-hydro-mechanical formulation, the following equations are needed: i. solid and fluid phase mass balance to derive a flow equation, ii. bacteria mass balance to derive a transport equation and iii. total momentum balance.

Mass balance equations and momentum balance equation are given for solid andfluid phases, bacteria and mixturerespectively:

$$\frac{\partial}{\partial t}\big((1-n)\rho^s\big) = -\boldsymbol{\nabla}\cdot\big((1-n)\rho^s\boldsymbol{V}^s\big) + k_{att}(nC) - k_{det}(\rho^b S) + k_g(\rho^b S) - k_d(\rho^b S) \quad (1)$$

$$\frac{\partial}{\partial t}(n\rho^f) = -\boldsymbol{\nabla}\cdot(n\rho^f\boldsymbol{V}^f) - k_{att}(nC) + k_{det}(\rho^b S) + k_g(nC) - k_d(nC)$$

$$\frac{\partial}{\partial t}(nC+\rho^b S) = -\boldsymbol{\nabla}\cdot\big((nC+\rho^b S)\boldsymbol{V}^s\big) - \boldsymbol{\nabla}\cdot\Big(n(C\boldsymbol{V}^f - \boldsymbol{D}\boldsymbol{\nabla}C)\Big) + \big(k_g - k_d\big)(nC+\rho^b S) \quad (2)$$

$$\boldsymbol{\nabla}\cdot\boldsymbol{\sigma} + \rho\boldsymbol{F} = 0 \quad (3)$$

where n the porosity, ρ^s, ρ^f the solid and fluid phase density, S the attached bacteria density, C the suspended bacteria concentration, k_{att}, k_{det} the attachment and detachment rates of bacteria, k_g, k_d the growth and decay rates of bacteria, $\boldsymbol{V}^s$, $\boldsymbol{V}^f$ the macroscopic velocity of the solid and fluid phases, $\boldsymbol{D}$ the hydrodynamic dispersion tensor, $\boldsymbol{\sigma}$ the volume averaged total stress and $\boldsymbol{F}$ the total body force per unit volume of porous medium at a macroscopic point.

3.2 Constitutive Equations

The balance equations introduced so far are independent of material properties. Constitutive conditions are needed to complete the description of the material behaviour.

The mechanical constitutive law is developed as followsand the relative fluid-solid velocity is governed by Darcy's law:

$$\boldsymbol{\sigma}' = \boldsymbol{D}^e : \boldsymbol{\varepsilon}^e = \boldsymbol{D}^e : (\boldsymbol{\varepsilon}^e_{mec} + \boldsymbol{\varepsilon}^e_{bio}) \;\; with \;\; d\boldsymbol{\varepsilon}^e_{bio} = -\alpha_{bio} dS\boldsymbol{I} \quad (4)$$

$$n(\boldsymbol{V}^f - \boldsymbol{V}^s) = -K(\boldsymbol{\nabla}p^f + \rho^f\boldsymbol{g}) \quad (5)$$

where $\boldsymbol{D^e}$ the elastic tensor, $\boldsymbol{\varepsilon^e}$ the solid elastic strain with a mechanical part $\boldsymbol{\varepsilon}^e_{mec}$ and a biological part $\boldsymbol{\varepsilon}^e_{bio}$, α_{bio} the biological expansion coefficient of the solid grains due to the attachment of bacteria and K the permeability.

3.3 Field Equations

The field equations are obtained by combining the mass balance equations, introducing the related constitutive equations and considering a linear equilibrium isotherm between S and C, i.e. $S = K_d C$ with $K_d = n/\rho^b * k_{att}/k_{det.}$

Upon combining mass balance equations for solid and fluid phases, considering the fluid compressibility $\partial_t \rho^f = \rho^f \beta_f \partial_t p^f$ and introducing Darcy's law, the flow equation is obtained:

$$\nabla \cdot \boldsymbol{V}^s - \nabla \cdot \left(K(\nabla p^f + \rho^f \boldsymbol{g})\right) + n\beta_f \frac{\partial p^f}{\partial t} - (k_g - k_d)\left(\frac{1}{\rho^f} n + \frac{1}{\rho^s} \rho^b K_d\right) C = 0 \quad (6)$$

Mass balance of bacteria combined with Darcy's law yield the transport equation:

$$\frac{\partial}{\partial t}\left(\left(1 + \frac{\rho^b}{n} K_d\right) C\right) + \nabla \cdot \left(\left(1 + \frac{\rho^b}{n} K_d\right) C \boldsymbol{V}^s\right) - \nabla \cdot \left(C \frac{K}{n} (\nabla p^f + \rho^f \boldsymbol{g})\right)$$

$$-\nabla \cdot (\boldsymbol{D} \nabla C) - (k_g - k_d)\left(1 + \frac{\rho^b}{n} K_d\right) C = 0 \quad (7)$$

Considering Terzaghi's effective stress, the following equation accounting for the hydro-mechanical coupling is derived from momentum balance of the mixture:

$$\nabla(\boldsymbol{D^e} : (\boldsymbol{\varepsilon}^e_{mec} + \boldsymbol{\varepsilon}^e_{bio}) - p^f \boldsymbol{I}) + \rho \boldsymbol{g} = 0 \quad (8)$$

The final system is composed of three equations (6), (7) and (8) that determine the three principal unknowns. It is implemented in the finite element code LAGAMINE [7,8].

4 NUMERICAL RESULTS

To test the validity of the developed biogrout transport formulation, numerical simulations of column injection tests are carried out in a vertical straight pipe (Figure 2). Results are given for a drained column with free concentration at its top under rapid injection (increase of concentration from 0 to 1 kg/m^3 during 10 s followed by constant injection during 2000 s). The column is considered packed with Jossigny silt [9,10].

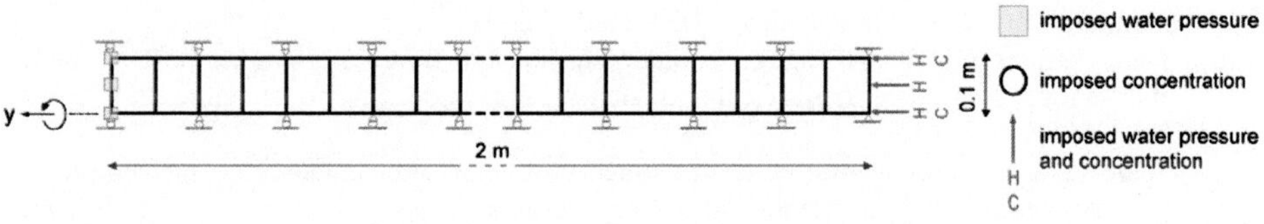

Fig. 2. Column used for injection tests

Figure 3a presents the evolution of the front of concentration along the column. It can be noticed that the porosity profile (Figure 3b) is symmetrical to the concentration profile, translating its bio-affected nature. Biologically induced stresses are presented in Figure 3c. As the lateral dilatation is prevented and a bio-elastic law is considered, an increase in biogrout concentration leads to an increase of lateral compressive stresses.

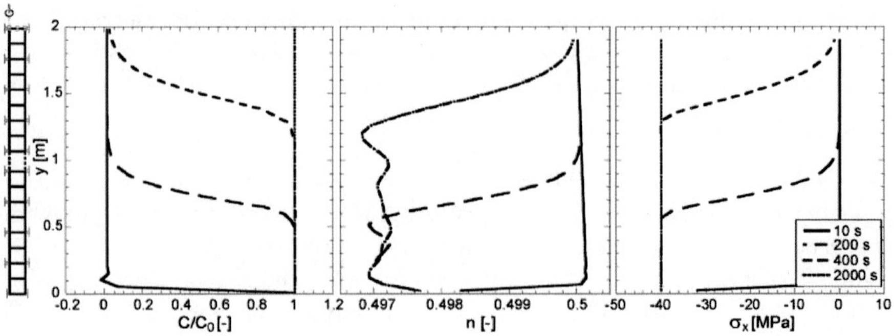

Fig. 3. (a) Biogrout concentration, (b) porosity and (c) biologically induced stresses profiles

5 CONCLUSIONS

Field equations have been developed for the case of propagation of miscible and reactive biogrout injection in a saturated and deformable porous medium. The numerical implementation of the formulation has been carried out in the finite element code LAGAMINE. Finite element modelling of column injection tests validate the formulation and demonstrate the potentiality of the developed model. The simulation results emphasize the significant role of bio-mechanical processes in the global MICP response of the materials.

REFERENCES

[1] Li, B.L., Loehle, C., Malon, D.: Microbial transport through heterogeneous porous media: Random walk, fractal and percolation approaches. Ecological Modelling 85, 285–302 (1996)

[2] Sen, T.K., et al.: Bacterial transport in porous media: New aspects of the mathematical model. Colloids and Surfaces A: Physicochemical and Engineering Aspects 260, 53–62 (2005)

[3] Tufkenkji, N.: Modeling Microbial Transport in porous media: Traditional approaches and recent development. Advances in Water Resources 30, 1455–1469 (2007)

[4] Bouchelaghem, F., et al.: Real scale miscible grout injection experiment and performance of advection-dispersion-filtration model. International Journal for Numerical and Analytical Methods in Geomechanics 25, 1149–1173 (2001)

[5] Chupin, O., Saiyouri, N., Hicher, P.Y.: Modeling of a semi-real injection test in sand. Computers and Geotechnics 36(6), 1039–1048 (2009)

[6] Saada, Z., et al.: Modelling of cement suspension flow in granular porous media. International Journal for Numerical and Analytical Methods in Geomechanics 29(7), 691–711 (2005)

[7] Charlier, R.: Approche unifiée de quelques problèmes non linéaires de mécanique des milieux continus par la méthode des élémentsfinis. Ph.D. Thesis, Université de Liège (1987)

[8] Collin, F.: Couplages thermo-hydro-mécaniques dans les sols et les roches tendres partiellement saturés. Ph.D. Thesis, Université de Liège (2003)

[9] Nuth, M.: Constitutive modelling of unsaturated soils with hydro-geomechanical couplings. Ph.D. Thesis, Swiss Federal Institute of Technology of Lausanne (2009)

[10] Cui, Y., Delage, P.: Yielding and plastic behaviour of an unsaturated compacted silt. Géotechnique 46(2), 291–311 (2009)

INTERPRETING THE "COLLAPSING" BEHAVIOR OF UNSATURATED SOIL BY EFFECTIVE STRESS PRINCIPLE

Ning Lu

Division of Engineering
Colorado School of Mines
1700 Illinois Street, Golden, Co, 80401
e-mail: ninglu@mines.edu web page: http://inside.mines.edu/~ninglu/

Summary. This paper examines the applicability of suction stress-based effective stress in describing the "collapsing" behavior of soil in oedometer testing under inundation conditions. The experimental data, originally used to invalidate Bishop's effective stress in the earlier sixties, are used to illustrate that the suction stress-based effective stress can well reconcile the phenomenon of soil sample "collapsing" during inundation process. It is shown that during inundation, suction stress increases drastically, leading to drastic reduction in effective stress and the failure of soil sample in oedometer without any changes in total stress.

Keywords: consolidation, effective stress, suction stress, unsaturated soil.

1 INTRODUCTION

In recent years, the suction stress characteristic curve (SSCC) has been introduced to represent the state of stress in unsaturated soil [1, 2]. In a broad sense, it is an expansion and extension of both Terzaghi's effective stress for saturated soil (e.g., [3]) and Bishop's effective stress for unsaturated soil [4]. Like previous effective stress approaches, the suction stress approach seeks a single stress variable that is responsible for the mechanical behavior of earth materials. However, suction stress differs from Terzaghi's "skeleton" stress in that forces contributing to suction stress are selfbalanced at the inter-particle level and thus do not originate from total stress that passes on from one particle to another. Suction stress originates from the available interaction energy at the soil solid surface that can be conceptualized to exist in the forms of van der Waals and double-layer forces, surface tension, and solid-liquid interface forces due to pore water pressure. A macroscopic continuum representation of suction stress is the isotropic tensile stress that can be simply determined from uniaxial tensile strength tests of unsaturated earth materials and other granular materials. The suction stress conception differs from Bishop's effective stress (i.e., $\sigma' = (\sigma - u_a) + \chi(u_a - u_w)$) mainly in that there is no need to define the coefficient of effective stress χ and it can capture changes in effective stress when soil is dry.

2 SUCTION STRESS-BASED EFFECTIVE STRESS

The effective stress principle, under the framework of the SSCC, can be expressed as [2]:

R.I. Borja (Ed.): Multiscale and Multiphysics Processes in Geomechanics, SSGG, pp. 81–84.
springerlink.com

$$\sigma' = (\sigma - u_a) - \sigma^s \tag{1}$$

where u_a is the pore air pressure, σ is the total stress, σ' is the effective stress, and σ^s is defined as the SSCC of the soil with a general functional form of [1]

$$\sigma^s = \begin{cases} -(u_a - u_w) & u_a - u_w \leq 0 \\ -(u_a - u_w) \Big/ \left\{1 + [\alpha(u_a - u_w)]^n\right\}^{(n-1)/n} & u_a - u_w \geq 0 \end{cases} \tag{2}$$

where u_w is the pore water pressure and n and α are empirical fitting parameters of unsaturated soil properties with α being the inverse of air-entry pressure for water saturated soil and n being the pore-size distribution parameter. Lu and Likos [1, 2] showed that the suction stress characteristic curve, σ^s, could be obtained either by shear strength tests or by soil-water characteristic curve (SWCC). The SWCC models are traditionally used to quantify the relationship between the normalized volumetric water content or degree of saturation and matric suction. If van Genuchten's SWCC model [5] is used, the normalized degree of saturation can be expressed as

$$S_e = \frac{S - S_r}{1 - S_r} = \left\{\frac{1}{1 + [\alpha(u_a - u_w)]^n}\right\}^{(n-1)/n} \tag{3}$$

A closed-form expression for suction stress for the full range of saturation can be arrived at [1, 6]:

$$\sigma^s = -\frac{S_e}{\alpha}\left(S_e^{n/(1-n)} - 1\right)^{1/n} \tag{4}$$

With eq. (2), the closed-form equation for effective stress in the entire pore water pressure range is:

$$\sigma' = \begin{cases} \sigma - u_a + (u_a - u_w) & u_a - u_w \leq 0 \\ \sigma - u_a + (u_a - u_w) \Big/ \left\{1 + [\alpha(u_a - u_w)]^n\right\}^{(n-1)/n} & u_a - u_w \geq 0 \end{cases} \tag{5}$$

Eq. (5a) is Terzaghi's equation for saturated state, and eq. (5b) is for unsaturated state. It can be shown that eq. (5b) continuously and smoothly approaches eq. (5a) as matric suction approaches zero [6]. Suction stress, therefore, can be considered the effective stress in unsaturated soil under the condition of no external stress. For sandy soil, suction stress is zero at zero matric suction (saturated condition), and at some large value of matric suction, it reaches a minimum value at a given matric suction. The absolute minimum of suction stress for the typical sandy soil is in the order of several kPa. The down-and-up characteristic of suction stress can also be better illustrated by plotting eq. (5b). This behavior of minimum suction stress is well known for sand-sized granular media. It is important to note that this behavior cannot be effectively described using Bishop's effective stress.

3 JENNINGS AND BURLAND'S "COLLAPSING" EXPERIMENT REVISITED

Eq. (5) can reconcile some seemingly contradictory historic evidence against effective stress in unsaturated soils. For example, loess soils throughout the central and western part of the US can collapse under confining pressures of less than 70 kPa under near-saturated conditions [7]. In tests on a silt and a clay-sand mixture, Jennings and Burland [8] reported that upon inundation, specimens collapsed without any further change in the external loading. This experimental evidence has been used to question the utility of effective stress for unsaturated soil. Using Bishop's effective stress framework, Khalili et al. [9] offered a new explanation of the collapsing behavior with a proper elastoplastic conception. Because loess or clay-sand mixtures are likely to have highly non-linear SSCCs, inundation behind a sharp wetting front drastically and concurrently decreases matric suction and depletes suction stress leading to reduction in effective stress and shear failure absent a change in the total stresses. Jennings and Burland argued that if the effective stress decreases upon soaking, the specimen should expand or dilate, rather than collapse as observed and thus effective stress cannot explain the phenomenon of collapsing upon soaking in these soils. The suction stress-based effective stress, as illustrated here, can reconcile the "collapsing" behavior observed in oedometer tests. Jennings and Burland tested three different soils: silt, silty sand, and silty clay. Since they only provide a SWCC for the silt, we will re-interpret their results from oedometer tests on the air-dried silt. Fig. 1a is a plot of void ratio versus applied pressure for the silt under complete soaking (solid line), and under air-dry loading followed by soaking at different consolidation pressures (lines and symbols). Upon soaking, all curves converge to the completely soaked line as the soils collapse to a void ratio of about that of the completely soaked tests. Fig. 1b shows the measured SWCC for the silt, from which we estimated the parameters α and n. These parameters were then used to determine the suction stress characteristic curve by eq. (4), also shown in Fig. 1b. For all specimens A–D, the initial compression is at the air-dry state, which leads to suction stress of about -350 kPa. Upon soaking, suction stress follows the arrows along the SSCC and increases drastically to near zero at 100% saturation leading to a reduction in effective stress of about 350 kPa as predicted by eq. (5).

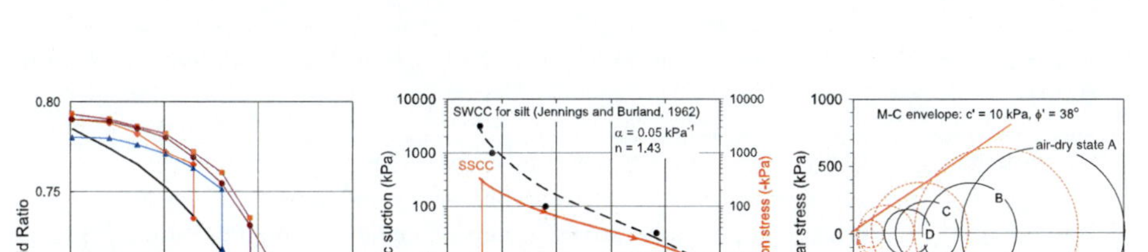

Fig. 1. Re-interpretation of the collapsing tests by Jennings and Burland [8]: (a) silt compression soaking tests, (a) measured soil water characteristic curve for the silt and predicted suction stress characteristic curve by eq. (5), (c) state of stress for the silt tests at the collapse state (shear failure) in the shear stress-effectives stress space.

Re-analyzing the state of stress in shear stress-effective stress space, we show that the state of stress for all specimens A–D in Fig. 1c as the solid circles prior to soaking. In plotting Fig. 1c, we assume that the silt has a drained cohesion of 10 kPa and a friction angle of 38°, and a lateral stress coefficient of $1 - \sin\phi'$ in the oedometer. Upon soaking, the effective stresses would reduce to the applied-stress conditions, as shown by the dashed circles. It is evident from Fig. 1c that all circles exceed the limit of the Mohr-Coulomb envelope, indicating that failure occurs in all these cases. It is important to note that because the soaking process creates sharp wetting fronts in all cases, the failure at the grain level is not uni-directional and along one failure plane as it is in the traditional shear and triaxial tests, it is multi-directional and at every point at the wetting front leading to the collapse of the fabric progressively throughout the entire specimen as the void ratio converges to the complete soaked compression line. As suggested in [9], the post-failure behavior of these specimen can be completely described under the effective stress framework with a proper plasticity model. Jennings and Burland restrained their interpretation of the state of stress and volumetric changes prior to the failure, thus entirely overlooking the failure and post failure behavior. Therefore, we conclude that the collapsing behavior reported in the soaking tests by Jennings and Burland can be fully reconciled using the effective stress framework.

4 CONCLUSIONS

It is shown that the original oedometer tests data, used by Jennings and Burland [8] to invalidate the Bishop's effective stress, was misinterpreted. If the suction stress-based effective stress is used, the "collapsing" phenomenon can be explained and reconciled by the effective stress principle. Upon inundation, soil samples in the oedometer drastically increase in suction stress, leading to drastic decreases in effective stress and shear failure.

REFERENCES

[1] Lu, N., Likos, W.J.: Unsaturated Soil Mechanics, 556 p. John Wiley and Sons, New York (2004)
[2] Lu, N., Likos, W.J.: Suction stress characteristic curve for unsaturated soil. Journal of Geotechnical and Geoenvironmental Engineering 132(2), 131–142 (2006)
[3] Terzaghi, K.: Theoretical Soil Mechanics, 510 p. John Wiley and Sons Inc., New York (1943)
[4] Bishop, A.W.: The principle of effective stress, Teknisk Ukeblad I Samarbeide Med Teknikk, Oslo, Norway, vol. 106, pp. 859–863 (1959)
[5] van Genuchten, M.T.: A closed-form equation for predicting the hydraulic conductivity of unsaturated soils. Soil Science Society of America Journal 44, 892–898 (1980)
[6] Lu, N., Godt, J., Wu, D.: A closed form equation for effective stress in unsaturated soil. Water Resources Research (2010), doi:10.1029/2009WR008646
[7] Holtz, W.G., Gibbs, H.J.: Consolidation and related properties of loessial soils. In: Special Technical Publication, vol. 126, pp. 9–26. ASTM, Philadelphia (1951)
[8] Jennings, J.E.B., Burland, J.B.: Limitation to the use of effective stresses in unsaturated soils. Géotechnique 12, 125–144 (1962)
[9] Khalili, N., Geiser, F., Blight, G.E.: Effective stress in unsaturated soils: review with new evidence. International Journal of Geomechanics 4, 115–126 (2004)

COUPLED ANALYSIS OF DOUBLE POROSITY SWELLING CLAYS

Antonio Gens[1,*], Leonardo do N. Guimarães[2], Marcelo Sánchez[3], and Beatriz Vallején[1]

[1] Department of Geotechnical Engineering and Geosciences
Universitat Politècnica de Catalunya
Jordi Girona 1-3, Barcelona Spain
* e-mail: antonio.gens@upc.edu, web page: http://www.upc.edu
[2] Departamento de Engenharia Civil
Universidade Federal de Pernambuco,
Rua Acadêmico Hélio Ramos, s/n. Cidade Universitária, Recife, Brazil
e-mail: leonardo@ufpe.br, web page: http://www.ufpe.br
[3] Department of Civil Engineering
Texas A&M University
College Station, Texas, USA
e-mail: msanchez@civil.tamu.edu, web page: http://www.tamu.edu/

Summary. The understanding of the coupled behaviour of highly active swelling clays is better achieved considering two structural levels and their interactions in the context of a double-porosity model. Hydraulic equilibrium between the two porosity levels is not assumed. Applications to hydration swelling tests of bentonite powder-pellets mixtures and to the chemomechanical behaviour of bentonites are presented.

Keywords: chemomechanics, coupled processes, double porosity, swelling clays.

1 INTRODUCTION

Swelling clays are being intensively studied as candidate materials for engineered barriers and seals in the storage and disposal of high-level nuclear waste in deep geological repositories. Often the behaviour of swelling clays can be better understood by considering two structural levels: microstructural and macrostructural as well as their interactions [1]. This dual behaviour is reinforced when the material is composed of a mixture of powder and highly compacted pellets (Figure 1a), an attractive material because a relatively high value of dry density can be achieved with minimal compaction effort. A double-porosity formulation has been developed that together with a generalised plasticity constitutive law [2], can be used to reproduce the coupled behaviour of this type of materials. Two applications are briefly presented herein: hydration of a mixture of bentonite powder and pellets and the mechanical response of a bentonite to chemical changes.

2 FORMULATION

The most important feature of the formulation is the assumption that there may be no equilibrium between the two porosity levels leading thereby to an exchange of water between them (Figure 1b). Water exchange is simply described by

R.I. Borja (Ed.): Multiscale and Multiphysics Processes in Geomechanics, SSGG, pp. 85–88.
springerlink.com

$$\Gamma^w = \gamma(\psi_M - \psi_m) \tag{1}$$

where Γ^w is the water exchange term, γ is a parameter (usually called the leakage parameter) and ψ is the total water potential. It is assumed that only matric and gravitational potential contribute to the total potential of the macrostructure but an additional osmotic component may also contribute to the microstructural potential [3]. As the water exchange takes place at the same point in the domain, water exchange is therefore driven by suction differences alone.

Using the concept of material derivative, the balance equation for the solid phase can be written as

$$\frac{D\phi}{Dt} = \frac{D\phi_M}{Dt} + \frac{D\phi_m}{Dt} = \frac{(1-\phi)}{\rho_s}\frac{D\rho_s}{Dt} + (1-\phi_M - \phi_m)(\dot{\varepsilon}_{vM} + \dot{\varepsilon}_{vm}) \tag{2}$$

where D denotes material derivative, ρ_s is the solid density, $\dot{\varepsilon}_v$ is the volumetric strain increment and t is time. Capital M denotes macrostructure, lowercase m microstructure. Then the water mass balance equation becomes

$$\frac{\partial}{\partial t}(S_{lj}\phi_j) + \nabla \cdot (\mathbf{j}_{lj}) \pm \Gamma^w = f_j^w \; ; \; j = M, m \tag{3}$$

where S_{lj} is the liquid saturation of medium j, $\mathbf{j}_{lj}$ is the total mass fluxes of water in the liquid phase and f_j^W is the external mass supply of water per unit volume in medium j.

This balance equation must be coupled to the momentum balance equation and suitable constitutive laws to close the formulation for analysis.

a)

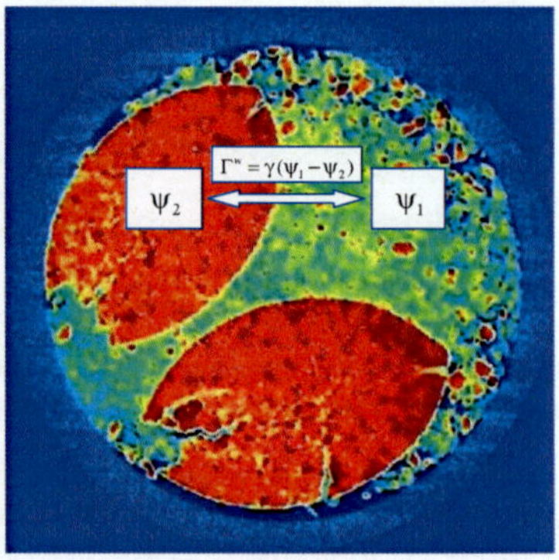

b)

Fig. 1. a) Mixture of bentonite pellets and bentonite powder used as sealing material in an experimental shaft b) Scheme of a double-porosity material with a water exchange term

3 HYDRATION OF A MIXTURE OF BENTONITE POWDER AND PELLETS

A number of high-quality swelling pressure tests were performed on mixtures of bentonite powder and highly compacted pellets in the CEA (France) and CIEMAT (Spain) laboratories [4]. Different mixture dry densities were used. A rather complex behaviour was noted involving size effects and non-monotonic development of swelling pressures. Microstructural observations were useful to identify the two porosity levels in the material. The formulation proved capable to reproduce satisfactorily the observed behaviour (Figures 2, 3). It has also been successfully applied to a full-scale test involving the sealing of an experimental shaft.

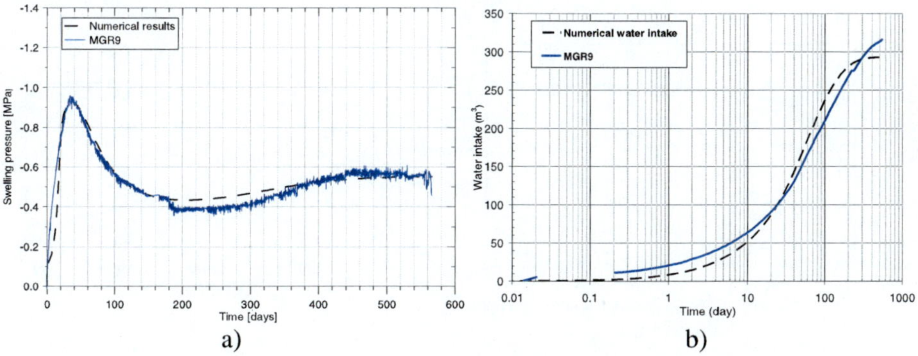

Fig. 2. Observed and computed values for a test on a sample of 1.45 g/cm^3 dry density: a) Swelling pressure evolution. b) Water intake evolution

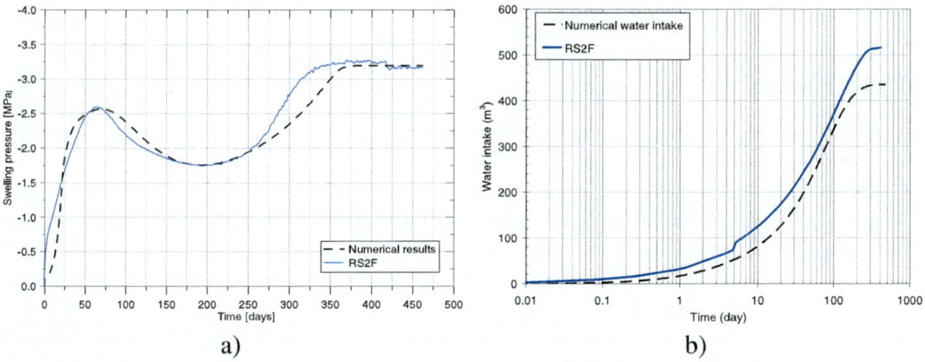

Fig. 3. Observed and computed values for a test on a sample of 1.60 g/cm^3 dry density: a) Swelling pressure evolution. b) Water intake evolution

4 CHEMOMECHANICAL BEHAVIOUR OF BENTONITE

The chemical environment may affect significantly the mechanical behaviour of very active clays such as bentonites. The double-structure model is especially suited to incorporate geochemical effects because they can be ascribed directly to the microstructure, the location where the relevant phenomena actually take place [3]. To illustrate the performance of the formulation, a chemical consolidation test described in [5] has been modelled using a fully coupled chemo-hydro-mechanical formulation [6]. The material tested is sodium bentonite and the samples were prepared from slurry with an initial void ratio of 4.6. In the test, the sample can drain only from the top, whereas pore pressure is measured at the bottom. The sample is subjected to a load of 100 kPa first. Once mechanical consolidation is finished, the specimen is placed in contact with a KCl saline solution of 4.0M concentration through the upper surface of the sample. It can be observed in Figure 4 that the double-porosity formulation describes well the observed chemomechanical behaviour of the bentonite including the pore pressure development in time. It was also noted that a close reproduction of the observed results required the consideration of hydraulic non equilibrium between the two structural levels.

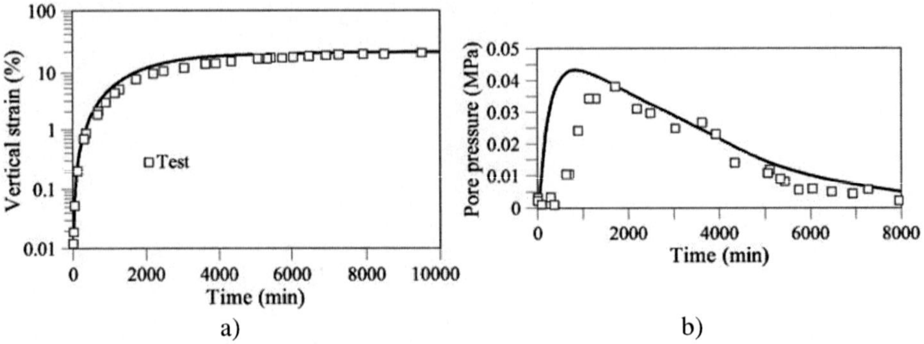

Fig. 4. Observed and computed results from a sodium bentonite oedometer sample exposed to a 4.0M solution of KCl: (a) vertical deformation; (b) pore pressure measured at the bottom of the specimen. Experimental results from [5]

REFERENCES

[1] Gens, A., Alonso, E.E.: A framework for the behaviour of unsaturated expansive clays. Canadian Geotechnical Journal 29, 1013–1032 (1992)

[2] Sánchez, M., Gens, A., Guimarães, L.N., Olivella, S.: A double structure generalized plasticity model for expansive materials. International Journal for Numerical and Analytical Methods in Geomechanics 29, 751–787 (2005)

[3] Gens, A.: Soil-environment interactions in geotechnical engineering. 47th Rankine Lecture.Géotechnique, 60, 3-74 (2010)

[4] Imbert, V., Villar, M.V.: Hydro-mechanical response of a bentonite pellets/powder mixture upon infiltration. Applied Clay Science 32, 197–209 (2006)

[5] Santamarina, J.C., Fam, M.: Changes in dielectric permittivity and shear wave velocity during concentration diffusion. Canadian Geotechnical Journal 32, 647–659 (1995)

[6] Guimarães, L.N., Gens, A., Olivella, S.: Coupled thermo-hydro-mechanical and chemical analysis of expansive clay subjected to heating and hydration. Transport in Porous Media 66, 341–372 (2007)

MULTIPHYSICS MODELING OF ELECTROKINETIC PHENOMENA IN UNSATURATED FINE–GRAINED SOILS

Cristina Jommi[1], Fabio Cattaneo[2], Guido Musso[3]
Claudio Tamagnini[4], and Diana Salciarini[4]

[1] Dipartimento di Ingegneria Strutturale
Politecnico di Milano, Italy
e-mail: jommi@stru.polimi.it
[2] Dipartimento di Ingegneria Civile, Architettura, Territorio e Ambiente
Università degli Studi di Brescia, Italy
e-mail: fcattaneo@stru.polimi.it
[3] Dipartimento di Ingegneria Strutturale e Geotecnica
Politecnico di Torino, Italy
e-mail: guido.musso@polito.it
[4] Dipartimento di Ingegneria Civile e Ambientale
Università degli Studi di Perugia, Italy
e-mail: tamag@unipg.it, diana@unipg.it

Summary. A multiphysics model for the analysis of coupled deformation and reactive flow of pore fluids, chemical species and electric current has been developed to assess the influence of the geochemical reactions taking place within the soil mass on the effectiveness of electrokinetic remediation treatments. The results of FE simulations have shown that the presence of carbonates in the solid phase may give rise to gas production, significant changes in the degree of saturation and buffering effects with respect to the pore water pH. Under such conditions, the efficiency of the electrokinetic treatment can be significantly reduced.

Keywords: electrokinetic phenomena, electrokinetic remediation, reactive transport and deformation.

1 INTRODUCTION

In the past twenty years, electrokinetic phenomena occurring in clayey soils upon the application of an electric field have been the subject of an intense research activity, aimed at exploring their potential in geoenvironmental applications, such as remediation of contaminated soils [1, 2] and contaminant diffusion control by means of active barriers for waste disposal facilities or electrokinetic fences for the control of contaminant plumes in polluted areas [3, 4]. All such applications rely crucially on the ability of the electric field to transport (ionic or non–polar) contaminants by means of electro-osmosis and ionic migration.

The effectiveness of environmental applications of electrokinetic phenomena—particularly when high current densities and long duration times are required—depends on the complex interactions between the transport phenomena in the pore fluid (of mass

R.I. Borja (Ed.): Multiscale and Multiphysics Processes in Geomechanics, SSGG, pp. 89–92.
springerlink.com

of pore fluids and ionic species and of the electric charge), the deformation of the solid skeleton and the geochemical reactions (precipitation, dissolution, sorption, aqueous phase reactions, electrolysis) taking place in the soil mass and at the electrodes. As pointed out by Acar and his coworkers [1, 2], the development of an acid front moving from the anode to the cathode due to water electrolysis is highly beneficial in electrokinetic extraction of contaminants in fine–grained soils, since it promotes the solubilization/desorption of contaminant species from the solid particles and increases the electrical conductivity of the soil.

Recent experimental observations by Airoldi et al. [5] on natural clay soils have shown that the presence of carbonates in the solid skeleton can have adverse effect on the electrokinetic treatment of a natural soil, mainly due to CO_2 generation—which reduces both electroosmotic and electrical conductivities—and the buffering effect with respect to pH changes. The aim of this work is to develop a multiphysics model for the analysis of coupled deformation and reactive flow of pore fluids, chemical species and electric current, which could be used to assess quantitatively the impact of carbonates geochemistry on the efficiency of electrokinetic remediation.

2 MULTIPHYSICS MODEL FOR LONG–DURATION EK PROCESSES

The governing equations of the multiphysics model for long-duration electrokinetic treatments considering reactive transport of carbonates are derived based on the assumptions of (i) deformable three-phase solid skeleton; (ii) incompressible solids; (iii) negligible streaming currents and electrophoresis; (iv) isothermal conditions. Assumption (i) represents an original contribution to the analysis of electrokinetic processes, which are typically modeled assuming fully saturated conditions and rigid solid skeleton.

In presence of carbonates, the chemical reactions taking place in the soil can be summarized as follows [6]:

$$\mathrm{H_2O} \leftrightarrow \mathrm{H^+} + \mathrm{OH^-} \tag{1}$$

$$\mathrm{CO_{2(g)}} + \mathrm{H_2O} \leftrightarrow \mathrm{H_2CO_3^*} \qquad \mathrm{H_2CO_3^*} = \mathrm{CO_{2(aq)}} + \mathrm{H_2CO_3} \tag{2}$$

$$\mathrm{H_2CO_3^*} \leftrightarrow \mathrm{H^+} + \mathrm{HCO_3^-} \qquad \mathrm{HCO_3^-} \leftrightarrow \mathrm{H^+} + \mathrm{CO_3^{-2}} \tag{3}$$

$$\mathrm{CaCO_3} \leftrightarrow \mathrm{Ca^{+2}} + \mathrm{CO_3^{-2}} \tag{4}$$

Equation (1) describes water auto-ionization; eq. (2) controls the solubility of CO_2 in the pore water as a function of its pressure, while eqs. (3) and (4) describe the dissociation of carbonic acid and the calcite dissolution/precipitation on the solid phase. The kinetics of the reactions taking place in the aqueous phase is assumed to be sufficiently fast to be considered almost instantaneous, with the exception of reaction (4), for which an appropriate kinetic velocity is adopted.

In a 1-dimensional setting, the balance of mass equations for pore water, pore gas and chemical species are given by

$$\frac{S_w n}{\rho_w}\frac{d\rho_w}{dt} + n\frac{dS_w}{dt} + S_w\frac{\partial v_x^s}{\partial x} + \frac{\partial w_x^w}{\partial x} = 0 \tag{5}$$

$$\frac{n(S_g + HS_w)}{\rho_g}\frac{d\rho_g}{dt} - \frac{nHS_w}{\rho_w}\frac{d\rho_w}{dt} + n\frac{dS_g}{dt} + S_g\frac{\partial v_x^s}{\partial x} + \frac{\partial w_x^g}{\partial x} = -R_{\mathrm{CO_2}}\frac{m_{\mathrm{CO_2}}}{\rho_g} \tag{6}$$

$$\frac{\partial(nc_i)}{\partial t} + \frac{\partial J_{x,i}}{\partial x} \pm R_i = 0 \tag{7}$$

$$\frac{\partial M_{\mathrm{CaCO_3}}}{\partial t} + R_{\mathrm{CaCO_3}}M_{\mathrm{CaCO_3}} = 0 \tag{8}$$

The term on the RHS of eq. (6) represents a source of gas production, proportional to the reaction rate of the reaction (2). Equation (8) provides the mass balance equation for the solid calcite; the quantity $R_{\mathrm{CaCO_3}}$ appearing in the second term represents the calcite production/consumption rate. The set of balance equations is completed by the balance of electric charge and momentum (in quasi–static conditions):

$$\frac{\partial j_x}{\partial x} + \frac{\partial Q_e}{\partial t} = 0\ ; \qquad \frac{\partial \sigma_x}{\partial x} + b_x = 0\ ; \qquad \sigma_x := \sigma_x'' - (S_w p_w + S_g p_g) \tag{9}$$

where Q_e is the soil-specific capacitance, and σ_x'' is the constitutive stress, linked to the skeleton deformation by the constitutive equation of the material.

Finally, the constitutive equations for the coupled flows of water, gas, chemical species and electric charge are given by

$$w_x^w = -\frac{1}{\mu_w}\,k_{w,\mathrm{rel}}\,\kappa\left(\frac{\partial p_w}{\partial x} - \rho_w b_x\right) - k_{e,\mathrm{rel}}\,\kappa_e\,\frac{\partial V}{\partial z} \tag{10}$$

$$\begin{aligned} w_x^g = &-\frac{1}{\mu_g}\,k_{g,\mathrm{rel}}\,\kappa\left(\frac{\partial p_g}{\partial x} - \rho_g b_x\right) \\ &+H\left\{-\frac{1}{\mu_w}\,k_{w,\mathrm{rel}}\,\kappa\left(\frac{\partial p_w}{\partial x} - \rho_w b_x\right) - k_{e,\mathrm{rel}}\,\kappa_e\,\frac{\partial V}{\partial x}\right\} \end{aligned} \tag{11}$$

$$J_{x,i} = -D_i^*\frac{\partial c_i}{\partial x} - z_i c_i u_i^* + c_i w_x^w \tag{12}$$

$$j_x = -\chi^*\frac{\partial V}{\partial x} \qquad \chi^* = \chi_s + \sum_{i=1}^{N} z_i F u_i^* c_i \tag{13}$$

where the hydraulic and electroosmotic conductivities are expressed by means of their intrinsic values (κ and κ_e) and the relative conductivities depending on the degree of saturation S_r; D_i^* and u_i^* are the effective diffusion coefficient and ionic mobility of species i, and χ^* is the effective electric conductivity, sum of a the superficial conductivity of the skeleton χ_s and of a Faraday term proportional to ionic mobility and charged species concentration.

3 NUMERICAL SIMULATIONS

The model described in the previous section has been implemented in the FE code COMSOL Multiphysics, and has been used to simulate an ideal 1–d electrokinetic filtration test on a natural silty clay. A constant p_w of 5 kPa above the reference gas

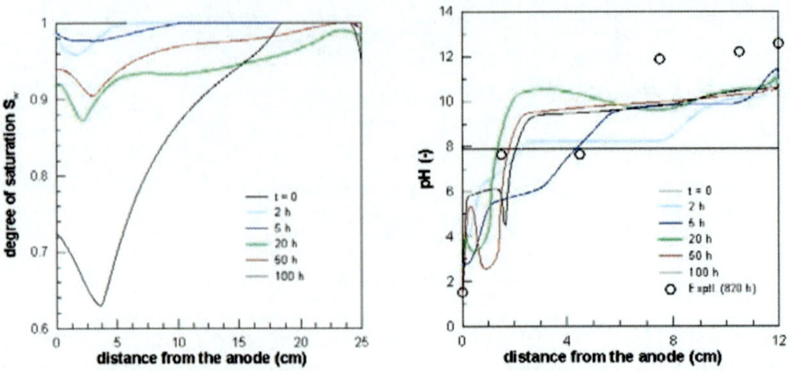

Fig. 1. Evolution with position and time of degree of saturation (left) and pore water pH (right).

pressure was imposed at both the anode and the cathode. Constant electric current density was imposed at the anode, while a reference null potential was assumed at the cathode. No inward flux was allowed for the chemical species at the anode. At the cathode the species were allowed to flow outwards by convection only. The measured value of pH was imposed at the anodic boundary. At the cathodic boundary, an inward flux of OH - was imposed, calculated on the basis of the current density and on the water autoionization. The model parameters were assigned – where possible – based on direct experimental information [5] or taken from the literature.

Some selected results of the numerical simulation are shown in Fig. 1. The calcite dissolution produces the development of gaseous CO_2 which induces a significant drop in the degree of saturation of the specimen. Moreover, the consumption of H^+ ions due to calcite dissolution prevents the motion of the pH front towards the cathode and maintains the pH close to the anode above 4.5÷5. This result is in agreement with the experimental data reported in [5].

REFERENCES

[1] Hamed, J., Acar, Y.B., Gale, R.: Pb(II) removal from kaolinite by electrokinetics. ASCE-JGGE 117, 241–271 (1991)

[2] Alshawabkeh, A.N., Acar, Y.B.: Electrokinetic remediation. II: Theoretical model. J. Geotech. Engng., ASCE 122, 186–196 (1996)

[3] Mitchell, J.K.: Conduction phenomena: From theory to geotechnical practice. Géotechnique 41, 229–340 (1991)

[4] Yeung, A.T., Sadek, S.M., Mitchell, J.K.: A new apparatus for the evaluation of electrokinetic processes in hazardous waste management. Geotech. Testing J. 15, 207–216 (1992)

[5] Airoldi, F., Jommi, C., Musso, G., Paglino, E.: Influence of calcite on the electrokinetic treatment of a natural clay. J. Applied Electrochemistry 39, 2227–2237 (2009)

[6] Appelo, C.A., Postma, D.: Geochemistry, groundwater and pollution. Balkema, Leiden (2005)

THE SCALE-UP PROBLEM FOR ION TRANSPORT IN KAOLINITE CLAYS INCLUDING PH-DEPENDENCE

Iury Igreja[1], Marcio A. Murad[1], Sidarta Lima[1], and Claude Boutin[2]

[1] Laboratório Nacional de Computação Científica LNCC/MCT
Av. Getúlio Vargas 333, 25651–070 Petrópolis, RJ, Brazil
e-mail: iuryhai@lncc.br, murad@lncc.br, sidarta@lncc.br, web page: http://www.lncc.br
[2] Univeristé de Lyon, Lab. Geomateriaux, DGCB URA CNRS 1652,
Ecole Nationale des Travaux Publics de l'Etat, 69518 Vaulx-en-Velin Cedex, France
e-mail: Claude.Boutin@entpe.fr, web page: http://www.entpe.fr

Summary. We propose a new multiscale computational model to simulate electrokinetic remediation processes in clayey soils. The macroscopic governing equations are derived using the homogenization procedure applied to the pore-scale description consisting of micro-pores saturated by an aqueous solution containing four monovalent ionic species (Na^+, H^+, Cl^-, OH^-) and charged solid particles surrounded by thin electrical double layers. The homogenized equations are discretized by the finite element method and applied to numerically simulate an electroosmosis experiment for decontamination of a clay sample by electrokinetics. Boundary conditions of Dankwerts type are postulated at the electrodes, which bring additional complexity to the iterative algorithm for solving the discrete model.

Keywords: clay, Dankcwerts' boundary conditions, electrokinetics, finite element, homogenization.

1 INTRODUCTION

In earlier papers [1, 2] some of the authors have developed new upscaling methods to describe electrochemical phenomena in active clays. Within the framework of homogenization-based approaches the complex effective response of the medium has incorporated pore-scale electro-chemo-mechanical couplings with phenomena such as electro-osmosis, electro-migration and osmotic swelling naturally appearing in the macroscopic forms of the convection-diffusion equations, Darcy's law and modified Terzaghis effective stress principle.

Despite the enormous improvement achieved by multiscale models, they were developed for electrolyte solutions composed of two fully dissociated monovalent ions Na^+ and Cl^- and constant pH=7. Consequently, the transport of H^+ and OH^- ions along with chemical protonation/deprotonation reactions between the H^+ ions and surface-charged groups lying on the clay particle surface leading to pH-dependency were neglected. First attempts at incorporating pH effects within the framework of homogenization were pursued by the authors in [3, 4]. Numerical approximations of the resultant multiscale model were carried-out and applied to computationally simulate the one-dimensional electro-osmotic experiment with Dirichlet boundary conditions enforced at the electrodes. In this work, we generalize

R.I. Borja (Ed.): Multiscale and Multiphysics Processes in Geomechanics, SSGG, pp. 93–96.
springerlink.com

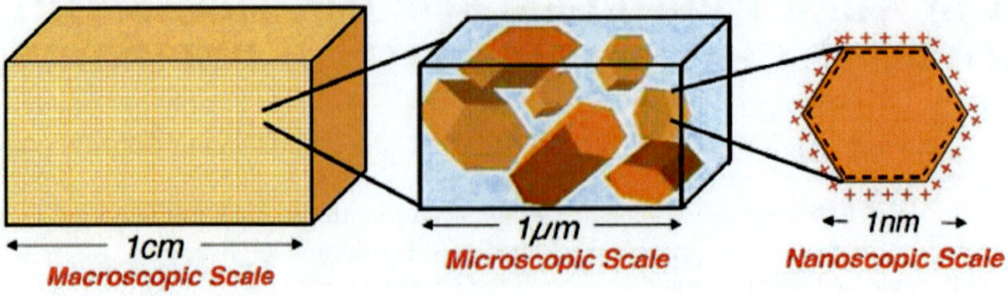

Fig. 1. Macroscopic, microscopic and nanoscopic portraits in kaolinite.

these simulations to incorporate more realistic boundary condition of Danckwerts type at the electrodes [5]. The assessment of the numerical results obtained within this new computational model is performed through a perturbation-based approach which gives rise to semi-analytical solutions of the electro-osmosis experiment capable of establishing accurate scenarios of electro-kinetic remediation of clayey soils.

2 REVIEW OF THE MULTISCALE MODEL

We begin by reviewing the three-scale electro-chemical model (see Figure 1) proposed in [3, 4] incorporating pH-effects. Let $\Omega = \Omega_s \cup \Omega_f \subset \mathbb{R}^3$ be the microscopic domain occupied by a biphasic porous media composed of clay particles and micro pores filled by a bulk fluid, and Γ_{fs} the solid / fluid interface. The hydrodynamics is governed by the classical Stokes problem

$$\nabla \cdot \mathbf{v} = 0 \quad ; \quad \mu_f \Delta \mathbf{v} - \nabla p = 0 \qquad \text{in } \Omega_f \tag{1}$$

where $\mathbf{v}$ is the fluid velocity, p the pressure and μ_f the water viscosity. The steady-state movement of the four ionic species with bulk concentrations C_{ib} ($i = Na^+$, H^+, Cl^-, OH^-) is governed by the Nernst-Planck equations

$$\begin{aligned} \nabla \cdot (C_{ib}\mathbf{v}) - \nabla \cdot \left[D_i \left(\nabla C_{ib} + C_{ib} \nabla \overline{\phi} \right) \right] + \dot{m} = 0, \\ \left(\dot{m} = 0 \ \text{ for } i = Na^+, Cl^- \right) \qquad \text{in } \Omega_f \end{aligned} \tag{2}$$

where D_i are the binary water-ion diffusion coefficients, $\overline{\phi} := F\phi/RT$ the dimensionless microscopic electrical potential with $\{F, R, T\}$ the set composed of Faraday constant, universal ideal gas constant and absolute temperature and $\dot{m}$ a source term that quantifies the mass production of H^+ and OH^- due to water hydrolysis.

To close our microscopic model it remains to postulate boundary conditions on the particle/micropore interface Γ_{fs}. Consider that the normal component of velocity and electro-diffusive ion fluxes vanish on Γ_{fs}. In addition, owing to the presence of the thin electrical double layer surrounding the particles we postulate the so-called Smoluchowsky slip [2] wherein the tangential velocity exhibits a discontinuity at the interface. Denoting $\boldsymbol{\tau}$ the unitary tangential vector at Γ_{fs} we have

$$\mathbf{v} \cdot \boldsymbol{\tau} = \widetilde{\epsilon}_0 \widetilde{\epsilon}_r \mu_f^{-1} \zeta \nabla \phi \cdot \boldsymbol{\tau} \qquad \text{on } \Gamma_{fs} \tag{3}$$

where $\widetilde{\epsilon}_0$ is the permittivity of the free space, $\widetilde{\epsilon}_r$ is the dielectric constant of the solvent and ζ the zeta potential. To close the microscopic model we need to build-up the

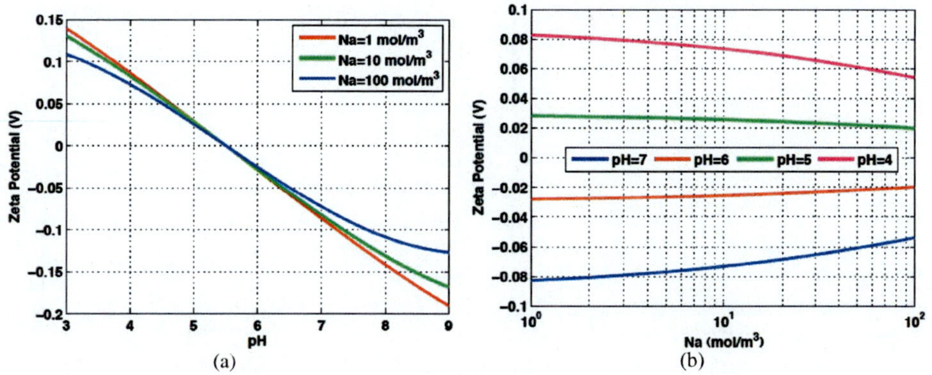

Fig. 2. Constitutive response of the ζ-potential as a function of pH and C_{Nab^+}.

constitutive law of the ζ-potential in terms of pH and salinity. To this end we make we proceed within the framework of the electrical double-layer theory with a nonlinear surface charge density due to protonation / deprotonation chemical reactions. This yields a single algebraic equation for the ζ-potential which has been solved in [3, 4].

The numerical constitutive response $\zeta = \zeta(C_{Nab^+}, C_{Hb^+})$ is depicted in Fig. 2. We may note strong dependence of the ζ-potential on the pH in a nearly skew-symmetric fashion around the point of zero charge (isoelectric point) $pH = 5.5$.

Application of the a formal homogenization procedure to the pore-scale model leads to the macroscopic equations [3]

$$\begin{cases} \nabla \cdot \mathbf{V_D} = \nabla \cdot [-K_P \nabla p - K_E \nabla \phi] = 0 \\ \nabla \cdot \mathbf{J_{Na^+}} = \nabla \cdot \left[C_{Nab^+} \mathbf{V_D} - D_{Na^+} \left(\nabla C_{Nab^+} + C_{Nab^+} \nabla \overline{\phi} \right) \right] = 0 \\ \nabla \cdot \mathbf{J_{H^+}} = \nabla \cdot \left[\Theta C_{Hb^+} \mathbf{V_D} - \widehat{D}_{H^+} \left(\nabla_x C_{Hb^+} + C_{Hb^+} \nabla \overline{\phi} \right) \right] = 0 \\ \nabla \cdot \mathbf{I_f} = \nabla \cdot \left[A \nabla C_{Nab^+} + B \nabla C_{Hb^+} + C \nabla \overline{\phi} \right] = 0 \end{cases} \tag{4}$$

where $\mathbf{V_D}$ the Darcy velocity, $\{K_P, K_E\}$ the hydraulic and electroosmotic permeability, $\{\mathbf{J_{Na^+}}, \mathbf{J_{H^+}}\}$ the convective/electro-diffusive ionic fluxes, $\{\Theta, \widehat{D}_{H^+}\}$ nonlinear coefficients given by $\Theta := 1 - K_W / C_{Hb^+}^2$ and $\widehat{D}_{H^+} := D_{H^+} + D_{OH^-} K_W / C_{Hb^+}^2$, $\mathbf{I_f}$ the electric current, C the electrical conductivity and $\{A, B\}$ the coupling coefficients for the current [3, 4]. The electroomsotic permeability is nothing but the average of the ζ-potential and therefore inherits its constitutive behavior.

3 APPLICATION TO AN ELECTROOSMOSIS EXPERIMENT

The macroscopic system is discretized by finite elements and applied to simulate an electroosmosis experiment. An anode and a cathode are placed at the positions $x = 0$ and $x = L$ respectively. Denoting $C_{Na^+}^{ref}$, $C_{H^+}^{ref}$ the sodium and H^+ concentration

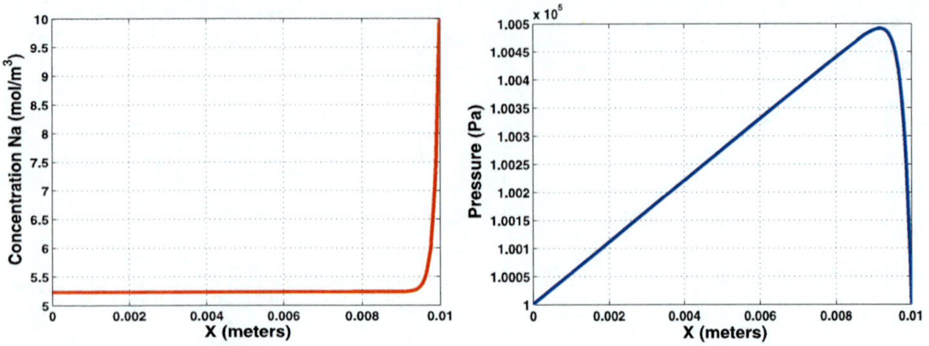

(a) (b)

Fig. 3. Sodium and pore pressure steady state profiles.

in the reservoir in the anode compartment and I_0 a given electric current postulate Danckwerts' boundary conditions at the elctrodes [5]

$$\mathbf{J_{Na^+}}|_{x=0} = C^{ref}_{Na^+}\mathbf{V_D} \qquad \mathbf{J_{Na^+}}|_{x=L} = C_{Na^+}\mathbf{V_D}$$
$$\mathbf{J_{H^+}}|_{x=0} = \Theta C^{ref}_{H^+}\mathbf{V_D} + I_0F^{-1} \qquad \mathbf{J_{H^+}}|_{x=L} = \Theta C_{H^+}\mathbf{V_D} + I_0F^{-1} \tag{5}$$

In Fig. 3 we display the sodium concentration and pressure profiles. In contrast to the case of Dirichlet boundary condition presented in [3], the enforcement of (5) leads to a plateau coupled with a sharp layer in the vicinity of the cathode (whose values can be estimated from perturbation analysis). In addition, the pressure gradient changes sign to fulfill the incrompressibility constraint. The more realistic simulation presented herein can be further explored to construct reliable electrokinetic remediation scenarios.

REFERENCES

[1] Moyne, C., Murad, M.: A two-scale model for coupled electro-chemo-mechanical phenomena and Onsager's reciprocity relations in expansive clays. Transport in Porous Media 62(3), 333–380, 63(1), 13–56 (2006)

[2] Murad, M.A., Moyne, C.: A dual-porosity model for ionic solute transport in expansive clays. Computational Geosciences 12, 47–82 (2008)

[3] Lima, S.A., Murad, M.A., Moyne, C., Stemmelen, D.: A three scale model for pH-dependent steady flows in 1:1 clays. Acta Geotechnica 3, 153–174 (2008)

[4] Lima, S.A., Murad, M.A., Moyne, C., Stemmelen, D., Boutin, C.: A three-scale model of pH-dependent flows and ion transport with equilibrium adsorption in kaolinite clays I and II. Transport in Porous Media (2010) (Online first)

[5] Alshawabkeh, A.N., Acar, Y.B.: Electrokinetic remediation: Theoretical model. J. Geotech. Eng. 122, 186–196 (1996)

FORMATION OF LAYERED FRACTURE AND OUTBURST BY PERCOLATION

X.B. Lu*, X.H. Zhang, and S.Y. Wang

Institute of Mechanics, Chinese Academy Sciences
*e-mail: xblu@imech.ac.ac

Summary. Formation mechanism of layered fracture and outburst in soils is investigated. It is shown that once the pore pressure is high enough to cause the over-cap failure, layered fracture or outburst occurs in liquefied or gasified sediments. The formation conditions are related to the strength of sediments and the boundary friction.

Keywords: gas hydrate, layered fracture, outburst.

1 INTRODUCTION

Layered fracture and outburst in soil sediment can occur in many cases, for example, during liquefaction of sand (also called water film or soil boiling) [1,2], during mining of coal, and during dissociation of gas hydrate [3,4].

Layered fracture in soils is a water or gas gap which forms when the excess pore pressure is larger than the weight and side friction at the upper part and accordingly water or gas is trapped in a thin gap between the upper part and the lower part [1]. Outburst is a kind of strong eruption of mixture of soils and water and gas [2].

If liquefiable or gasified sediment is overlain by a less permeable layer, then the overlying sediment (over-cap) can restrict pore water or gas to pass through. Once the pore pressure is larger than the weight and the side friction, the over-cap will be pushed upward to form a fracture or even crushed, which leads to a sudden unloading of pore pressure at the interface between the saturated or gasified sediment and the over-cap. Accordingly, an unloading wave will transmit into the sediment. Sediment can be fractured in layered form when the excess pressure is not too high; otherwise, the fracture failure becomes continuous and outburst occurs.

The main objective of this study is to investigate the mechanism and formation conditions of layered fracture and outburst in a soil layer. First, some typical experiments are introduced. Second, initiation conditions of these two types of failure forms are analyzed.

2 LAYERED FRACTURE

In general, layered fracture happens in a relatively low pore pressure. First, the over-cap is only pushed upward to form a gap by the pore fluid (gas or water or their mixture), and then one or more fractures occur in the liquefied or gasified zone if the pressure difference is high enough between the former fracture and the deeper zone.

R.I. Borja (Ed.): Multiscale and Multiphysics Processes in Geomechanics, SSGG, pp. 97–100.
springerlink.com

When the pressure p_0 in soil layer satisfies the condition

$$p_0 \geq p_a + \frac{\tau l}{A} L + \sigma_t \tag{1}$$

layered fracture initiates, in which p_a is the atmospheric pressure, p_0 is the pressure inside the soil, $\tau l/A$ is the side friction in unit length, and σ_t is the tensile strength.

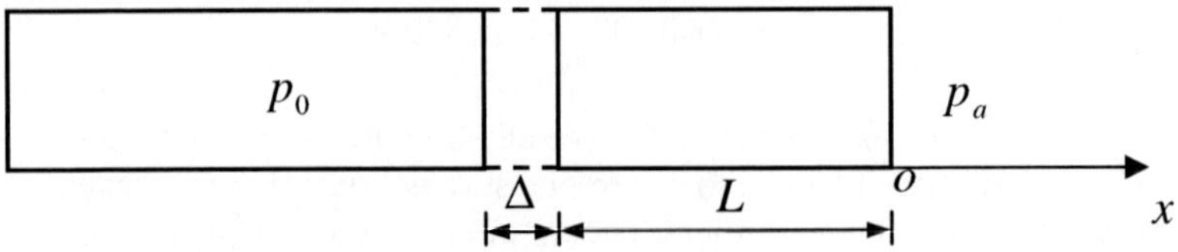

Fig. 1. Sketch of layered fracture

To analyze the movement of the layered zone, the following formulas are adopted:

$$p - p_a - \frac{\tau l}{A} L = \rho_s (1 - \varepsilon_0) L \cdot \ddot{\Delta} \tag{2}$$

with initial condition

$$t = 0, \Delta = \Delta_0, \dot{\Delta} = 0 \tag{3}$$

in which p is the pressure in the fracture, ρ_s is the density of soils, ε_0 is the initial porosity, Δ is the width of the fracture, and Δ_0 is the initial width of the fracture (Fig. 1).

With the movement of the fracture zone, the fracture expands, and the pressure inside it decreases, causing the pore fluid to seep into the fracture from the two sides. To analyze the development of the fracture width, the following equations are adopted. The fracture zone is denoted by zone I, and the soil layer after the fracture is denoted by zone II.

In zone I, the pressure p_1, density ρ_1, and velocity u_1 satisfy

$$\begin{cases} -\dfrac{\partial p_1}{\partial x} = \dfrac{\varepsilon_0 \mu}{k} u_1 \\ \dfrac{\partial \rho_1}{\partial t} + \dfrac{\partial \rho_1 u_1}{\partial x} = 0 \\ p_1 = \rho_1 a^2 \end{cases} \tag{4}$$

with initial condition

$$u_1 = 0, \ p_1 = p_0 \tag{5}$$

and boundary condition

$$x = -L, \ p_1 = p; \ x = 0, \ p_1 = p_a \tag{6}$$

In zone II, the pressure p_2, density ρ_2, and velocity u_2 satisfy

$$\begin{cases} -\dfrac{\partial p_2}{\partial x} = \dfrac{\varepsilon_0 \mu}{k} u_2 \\ \dfrac{\partial \rho_2}{\partial t} + \dfrac{\partial \rho_2 u_2}{\partial x} = 0 \\ p_2 = \rho_2 a^2 \end{cases} \tag{7}$$

with initial condition

$$u_2 = 0,\ p_2 = p_0 \tag{8}$$

and boundary condition

$$x = -\infty,\ p_2 = p_0;\ x = -L,\ p_2 = p \tag{9}$$

in which μ is the viscosity of pore fluid, k is the permeability, and a is the sonic speed.

The total flow rate Q from zone I and II into the fracture is

$$Q = Q_1 + Q_2 = \varepsilon_0 A(\rho_1 u_1 + \rho_2 u_2) \tag{10}$$

The mass and the density in the fracture are

$$m = m_0 + \int_0^t Q dt \tag{11}$$

$$\rho = \frac{m}{A \cdot \Delta} \tag{12}$$

in which m_0 is the initial mass of pore fluid in the fracture Δ_0, and $m_0 = \rho_0 A \Delta_0$.

Using the formula $p = \rho a^2$, the relation between the pressure inside the fracture and the fracture width is

$$p \cdot \Delta = a^2 \left(\int_0^t \frac{Q}{A} dt + \frac{m_0}{A} \right) \tag{13}$$

3 OUTBURST

Discontinuity theory is used here to analyze the initiation of outburst. The effective stress can be expressed as

$$\sigma_e = p - \sigma \tag{14}$$

in which σ_e is the effective stress, p is the pore pressure, and σ is the total stress.

In general, during the unloading of the total stress, the effective stress σ_e becomes gradually tensile. When the tensile stress reaches the maximum, the soil layer will include failure:

$$p - \sigma = \sigma_M - 2\nu[\sigma_s - p] \tag{15}$$

in which σ_M is the tensile strength of soil layer at $p=0$ and $\sigma_s=0$, ν is the Poisson's ratio, and σ_s is the lateral stress.

Assuming that outburst proceeds so fast that the relative movement between pore fluid and skeleton as well as the volume change of grains can be neglected, the relation between pressure and strain E is given by

$$p(\varepsilon_0+E)=p_0\varepsilon_0 \tag{16}$$

Before failure, the strain can be neglected, so that

$$\sigma=\sigma_D=\alpha p_0-\sigma_l \tag{17}$$

in which $\alpha=1-2\nu$, $\sigma_l=\sigma_M-2\nu\sigma_s$, and σ_l is the tensile strength, considering the effects of lateral pressure.

Using the C-J condition as in the case of steady deflagration, the criterion of outburst can be obtained as

$$p_0=\frac{2p_a+\sigma_l+\sqrt{(2p_a+\sigma_l)^2-4\alpha p_a^2}}{2\alpha} \tag{18}$$

If the side friction F_s is considered, then σ_D in the above equations is instituted by

$$\sigma_D^*=\sigma_D-F_s \tag{19}$$

4 CONCLUSIONS

It is shown that once the pore pressure is large enough to cause the failure of over-cap (forming a hole or being broken), layered fracture or outburst occurs in the liquefied or gasified sediments. Three critical pressures are present corresponding to the stable critical value, layer fracture critical value, and outburst critical value. The critical values are close to the values of the strength of sediments and the side friction.

REFERENCES

[1] Lu, X.B., Wu, Y.R., Zheng, Z.M.: Formation mechanism of cracks in saturated sand. Acta Mech. Sinica. 22(4), 377–383 (2006)

[2] Brennan, A.J., Madabhushi, S.P.G.: Liquefaction and drainage in stratified soil. ASCE J. Geotech. Geoenviron. Engrg. 131(7), 876–885 (2005)

[3] Yu, S.B., Cheng, C.M., Tan, Q.M., Ding, Y.S.: Damage of porous dedia containing pressurized gas by unloading and the maximum damage principle for critical outburst. Acta Mech. Sinica 29(6), 641–646 (1997) (in Chinese)

[4] Briaud, J.L., Chaouch, A.: Hydrate melting in soil around hot conductor. J. Geotech. and Geoenviron. Engrg. 123(7), 645–653 (1997)

PERMEABILITY EVOLUTION IN FRACTURED ROCK IN RESPONSE TO MECHANICAL AND CHEMICAL PROCESSES

Joseph P. Morris

Computational Geosciences Group
Lawrence Livermore National Laboratory
Now at:
Schlumberger – Doll Research
One Hampshire Street, MD-B408, Cambridge, MA 02139
e-mail: jpmorris@slb.com

Summary. Large-scale carbon capture and sequestration (CCS) projects are required to substantially reduce greenhouse gas emissions. The large rate and volume of injection induce pressure and stress gradients within the formation that could activate existing fractures and faults, or drive new fractures through the caprock. In addition, over longer periods of time, geochemical processes lead to evolution in the porosity and permeability of the storage domain. These processes involve coupling over multiple scales and require development of appropriate simulation capabilities. We present the results of investigations using two classes of computational tool: Distinct Element Methods and Boundary Element Methods. We present several demonstration applications of these codes to CO_2 injection scenarios. These examples highlight the advantages of explicitly including the geomechanical response of each interface within the formation.

Keywords: CO_2 sequestration, discrete-finite element method, fluid flow, fracture networks.

1 INTRODUCTION

The Livermore Distinct Element Code, LDEC, was developed by Morris et al. [1-4] as a combined distinct element–finite element (FE-DEM) code to simulate the deformation of extensive, jointed geologic rock masses. The DEM is naturally suited to simulating such systems because it can explicitly accommodate the blocky nature of natural rock masses. In addition, LDEC supports fully coupled fluid flow using both Smooth Particle Hydrodynamics (SPH) and unstructured fracture flow mesh methods [5]. While techniques such as LDEC can simulate extensive fractured masses, they must assume constitutive models for predicting the mechanical response of the individual fractures.

In contrast, it is also of interest to consider detailed coupling between various processes within single fractures. In practice, it is too computationally expensive to include the fine structures typical of a natural fracture (such as individual asperities) using finite elements or distinct elements. For this reason, a Boundary Element Method (BEM)-based code, Flex, was developed [6]. Using Flex, a fracture is approximated by potentially millions of asperities attached to elastic half-spaces that interact according to boundary element interaction terms. As the confining stress on the fracture changes, Flex uses the Fast Multipole Method to obtain a rapid prediction

R.I. Borja (Ed.): Multiscale and Multiphysics Processes in Geomechanics, SSGG, pp. 101–104.
springerlink.com

of the evolution in aperture within the fracture [6]. Subsequently, Flex-like capabilities were incorporated into a reactive transport code to become part of Frac-HMC (Fracture-based Hydro-Mechanical-Chemical processes), which has been used to investigate the relationship between dissolution and stress in fractures [7].

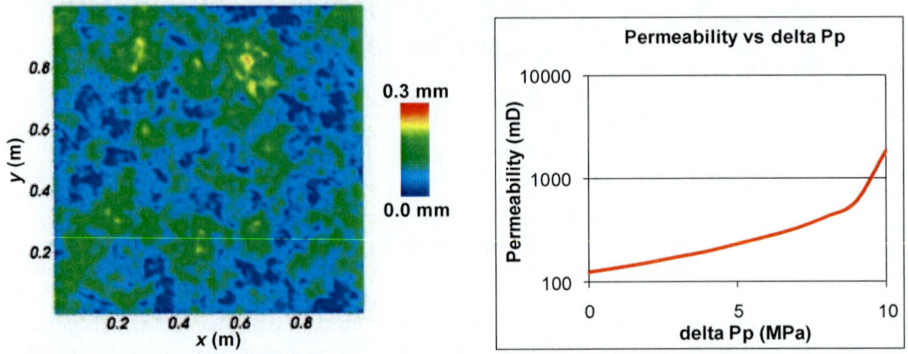

Fig. 1. Left: Aperture distribution within Frac-HMC representation of a single fracture. Right: Prediction of permeability evolution with increasing pore pressure. An order of magnitude increase in permeability is predicted as the pore pressure increases by 10 MPa.

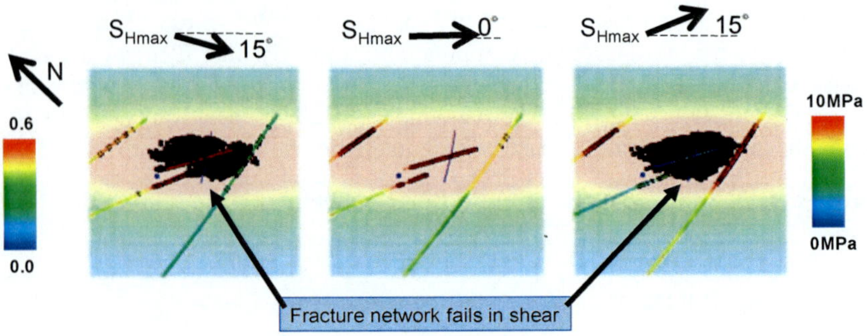

Fig. 2. In response to injection, portions of the faults and fractures experience shear slip depending upon the precise orientation of the in situ stress field. This implies that extensive permeability enhancement may be expected depending upon the conditions in the field.

2 APPLICATION OF LDEC AND FRAC-HMC TO THE IN SALAH PROJECT

The In Salah Joint Industry Project (a consortium consisting of BP, StatoilHydro and Sonatrach, hereafter referred to as the JIP) includes a CO_2 sequestration effort that has successfully injected millions of tons of CO_2 into a deep saline formation close to a producing gas field in Algeria. Here, we present predictions of permeability evolution associated with injection at In Salah using LDEC and Frac-HMC.

2.1 Prediction of Permeability Evolution during Fluid Injection Using Frac-HMC

This analysis predicts the evolution of permeability in response to normal displacement changes resulting from increases in pore pressure including details of aperture distribution within a fracture adjacent to an injector at In Salah. We constructed a model of a fracture within the reservoir based upon data provided by the In Salah JIP. The Flex-like capability within Frac-HMC calculates the change in aperture distribution throughout the fracture and associated permeability increase in response to increases in pore pressure. The results (see Figure 1) predict an order of magnitude increase in permeability due to hydromechanical deformation of pre-existing fractures during injection.

2.2 Simulations of Hydromechanical Response of Extensive Fracture Networks and Intersecting Faults Using LDEC

The previous section discussed the response of individual fractures to fluid injection. Further hydromechanical studies were conducted on the combination of fractures and faults within the reservoir. Figure 2 shows the response of the combined fracture and fault network to a hypothetical pore pressure increase as part of a parameter study (see [8] for more details). These calculations consider the poroelastic response of the fractured rock mass and include the redistribution of stresses through the combined fracture–fault network. The black regions highlight sections of the faults and fractures that fail and will provide enhanced permeability within the reservoir. We explored the effect of rotating the in situ stress relative to fractures/faults. This was intended to explore influence of uncertainty of in situ stress orientation, fracture strike variability and fault strike uncertainty/variability.

The results of these studies indicate that significant permeability evolution is expected within the reservoir due to combinations of fracture and fault deformation. One key observation is that the fracture network and in situ stress are oriented in such a way that the extent of permeability evolution is very sensitive to the orientation of the in situ stress field. In short, the shape and extent of the plume is predicted to be highly sensitive to small variations in the relative orientation to the in situ stress.

3 SIMULATIONS OF COMBINED CHEMICAL AND MECHANICAL PROCESSES IN A SINGLE FRACTURE USING FRAC-HMC

As was previously stated, Frac-HMC includes reactive transport capabilities [7]. Frac-HMC has been used to predict the evolution of aperture during dissolution within a fracture under varied stress conditions. For example, Figure 3 shows how the same fracture, under increasing confining stress, experiences different aperture evolutions due to dissolution. Such techniques are being used to understand the long-term evolution of permeability within CO_2 sequestration systems that include fractures and other interfaces.

4 CONCLUSIONS

We have described the features now implemented in the LDEC and Flex/Frac-HMC codes that allow us to simulate the evolving aperture distribution within a fractured rock mass. Such capabilities are now being used to predict site-specific performance of CO_2 sequestration sites due to combinations of mechanical and chemical effects. It is clear from the examples discussed that these approaches are complementary. While massively parallel capabilities such as LDEC can simulate extensive, fractured rock masses, they assume the existence of fracture-scale constitutive models. In contrast, techniques such as Flex/Frac-HMC predict permeability evolution including detailed

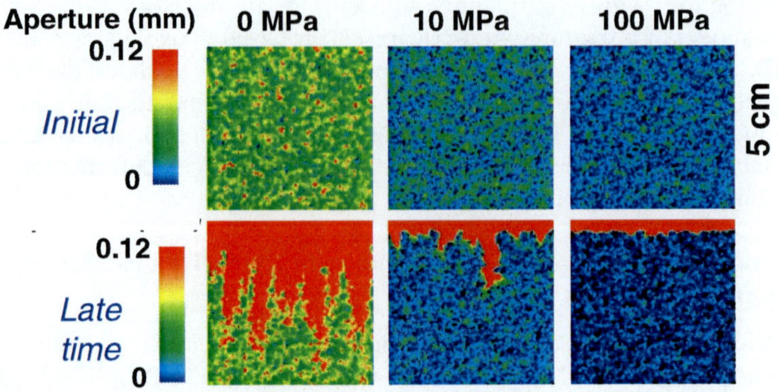

Fig. 3. Frac-HMC analysis of stress-dependent fracture aperture evolution. Initial aperture is shown at top and final aperture due to dissolution at bottom. Results toward the right correspond to increasing confining stress.

structure within individual fractures, but are typically more limited in terms of what stress boundary conditions can be accommodated.

ACKNOWLEDGMENTS

This study was performed under the auspices of the U.S. Department of Energy by Lawrence Livermore National Laboratory under Contract DE-AC52-07NA27344.

REFERENCES

[1] Morris, J.P., Glenn, L.A., Blair, S.C.: The distinct element method - application to structures in jointed rock. In: Meshfree Methods. LNCSE, pp. 291–306 (2002)

[2] Morris, J.P., et al.: Simulations of underground structures subjected to dynamic loading using the distinct element method. Engineering Computations 21, 348–408 (2004)

[3] Morris, J.P., et al.: Simulations of fracture and fragmentation of geologic materials using combined FEM/DEM analysis. International Journal of Impact Engineering 33, 463–473 (2006)

[4] Block, G., et al.: Simulations of dynamic crack propagation in brittle materials using nodal cohesive forces and continuum damage mechanics in the distinct element code LDEC. International Journal of Fracture 144, 131–147 (2007)

[5] Johnson, S.M., Morris, J.P.: Hydraulic fracturing mechanisms in carbon sequestration applications, In: Presented at 43rd U.S. Rock Mechanics Symposium & 4th U.S. - Canada Rock Mechanics Symposium, Asheville, North Carolina, June 28 - July 1 (2009)

[6] Pyrak-Nolte, L.J., Morris, J.P.: Single fractures under normal stress: The relation between fracture specific stiffness and fluid flow. Int. J. Rock Mech. And Min. Sci. 37, 245–262 (2000)

[7] Detwiler, R.L., Morris, J.P., Ezzedine, S.M.: Coupling micromechanics and fluid flow through discrete fracture networks: Quantifying effective permeability under variable stress conditions. Eos Transactions AGU 86(46), Fall Meeting Supplement, Abstract H13D-1433 (2006)

[8] Morris, J.P., Carroll, S.A., Mcnab, W.W., Hao, Y., Foxall, W.F., Ramirez, A.L.: Quarterly Report for Injection and Reservoir Hazard Management: The Role of Injection-Induced Mechanical Deformation and Geochemical Alteration at In Salah CO_2 Storage Project, LLNL-TR-412713 (March 2009)

SIMULATION OF HYDRATION AND ELASTIC PROPERTIES OF MONTMORILLONITE USING MOLECULAR DYNAMICS

Davoud Ebrahimi[1], Roland J.-M. Pellenq[1,2], and Andrew J. Whittle[1]

[1] Department of Civil and Environmental Engineering
Massachusetts Institute of Technology
e-mail: davebra@mit.edu, ajwhittl@mit.edu
[2] Centre Interdisciplinaire des Nanosciences de Marseille
Centre National de la Recherche Scientifique and Marseille Universite
e-mail: pellenq@cinam.univ-mrs.fr

Summary. This article describes molecular-level simulations of the adsorption isotherm and elastic properties for a Wyoming Na-montmorillonite using the General Utility Lattice Program. The selected clay mineral includes isomorphous substitutions in both the silica tetrahedral and octahedral gibbsite sheets. The resulting negative charges are balanced by Na+ cations. The authors have investigated the importance of different interatomic potentials in understanding the adsorption of water molecules (at ambient temperature, 300K). A unique combination of NPT and Grand Canonical ensemble simulations using Core–Shell potential shows intra-layer adsorption, while the same approach based on empirical CLAYFF potential produces only interlayer water adsorption. There is a correspondingly large difference in the theoretical elastic modulus normal to the clay sheets using the two different sets of force potentials. This article discusses the underlying reasons for these differences and reviews the experimental evidence supporting intra-layer water absorption in montmorillonite.

Keywords: clay mineral adsorption, clay swelling, mechanical properties, molecular simulation.

1 INTRODUCTION

Engineering properties of soil are highly affected by the clay content. Studying clay and its reaction with water is essential in understanding soil behavior. We want to model clay mineral and study changes in the fundamental elastic material properties of clay during hydration at the atomic scale. The existing literature on this subject has focused on swelling behavior of clay assuming rigid clay layers [1]. Measurements of the elastic properties of smectite clay are difficult because of the small grain size and high affinity for water [2, 3]; therefore, computational methods can provide a valuable tool for developing theoretical estimates of these properties.

2 COMPUTATIONAL METHOD

We have relaxed the rigid layer assumption using Core–Shell force field [4–6] that is widely used in the field of molecular simulation of mineral phases because it shows a high degree of transferability of parameters. In order to have a quantitative measure of sensitivity of our results with respect to the chosen force field, we have considered

R.I. Borja (Ed.): Multiscale and Multiphysics Processes in Geomechanics, SSGG, pp. 105–108.
springerlink.com

the effect of different force fields by simulating the system with CLAYFF [7] which uses partial charges rather than formal charges. We compute the elastic properties of clay minerals at different hydration states from the atomic-scale simulations. We use the General Utility Lattice Program (GULP) [8].

Using GULP, we can perform Molecular Dynamics (NVT or NPT - MD) and Grand Canonical Monte Carlo (GC - μ-V-T) simulations as well as calculating elastic properties which are of interest in our current research. We modeled clay hydration using the GCMC simulation at 300K temperature. This step provides the starting configuration at defined hydration states related to the experiment. Swelling and elastic deformation properties of clay are then calculated for the molecular unit cell in a series of MD simulations in isothermal–isobaric ensemble (NPT) at T=300K and P=0GPa.

The crystallography of 2:1 clay mineral has been taken from Refson et al. [9]. We are modeling $Na_{0.75}nH_2O[Si_{7.75}Al_{0.25}][Al_{3.5}Mg_{0.5}]O_{20}(OH)_4$, Wyoming Na-montmorillonite which carries isomorphous substitution in both tetrahedral and octahedral sheets.

3 RESULTS AND DISCUSSION

Figure 1a shows experimental and calculated equilibrium layer spacing at different levels of clay hydration using two force field parameters: Core–Shell and CLAYFF. Results of CLAYFF (circle) simulations show good agreement with experiment [10], while results using Core–Shell (diamond) potentials are far from experiment. Figure 2a, 2b shows equilibrated structure of the system at the production stage. Core–Shell force field indicates an important characteristic that has been missed in previous computer simulations about the structure and the position of water molecules in the system. The simulations show that water can be confined within the intra porosities of the octahedral and tetrahedral sheets, which build the structure of the clay mineral. Results of calculating elastic modulus perpendicular to the clay layers at different amount of water content are shown in Figure 1b. We performed one simulation using CLAYFF with starting configuration of having intra-layer water molecules (filled circle). Owing to the incompressibility of water molecules, existence of intra-layer molecules makes the clay–water system stiffer than the same configuration without them. The lower value of elastic modulus (around 20 GPa) corresponding to the CLAYFF simulations without intra-layer water molecules is close to the extrapolated values reported from acoustic measurements on cold-pressed samples of montmorillonite [2]. The upper value of elastic modulus (around 60GPa) corresponding to the case of having intra-layer water molecules is close to the values calculated from compressibility measurements for pyrophyllite [11] (by assuming poisson's ratio of 0.28 [3]) and obtained values from Brillouin scattering [12] on muscovite mica. The only direct experimental measurement of elastic modulus on clay mineral reported value of 6.2 GPa for dickite using atomic force acoustic microscopy [13] which is far from calulated values. These different values of the elastic modulus can be related to the correct structure and state of water molecules in the system.

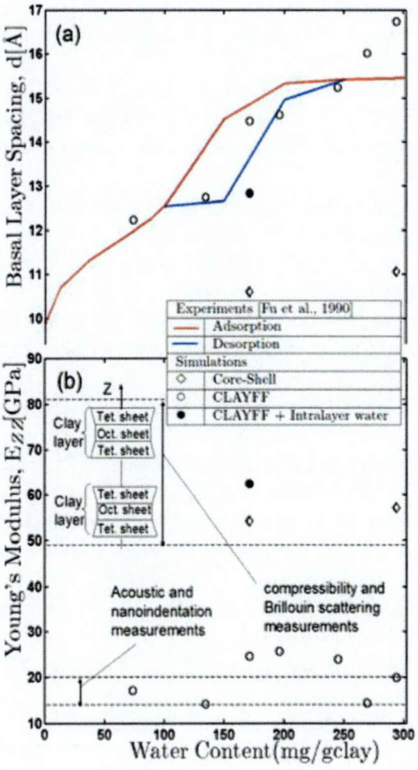

Fig. 1. Basal layer spacing and Young's modulus versus water content: (a) simulated and experimental swelling curves for Wyoming Na-montmorillonite, (b) calculated elastic modulus perpendicular to the clay layers

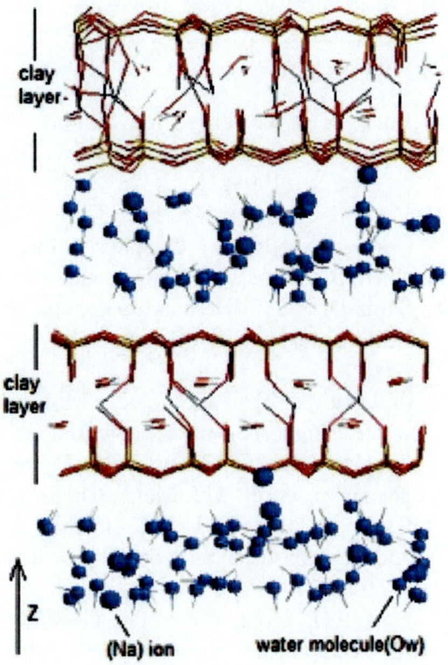

(a)

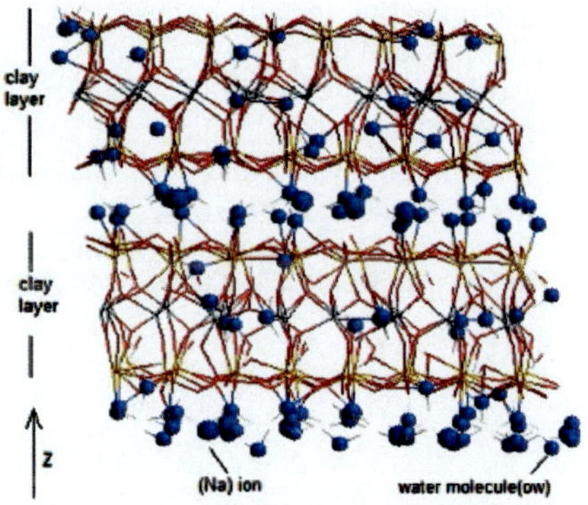

(b)

Fig. 2. An equilibrated configuration of the system containing 56 water molecules: (a) CLAYFF simulation, (b) Core–Shell simulation

ACKNOWLEDGMENT

This research was supported in part by the National Science Foundation through TeraGrid resources provided by NCSA and TACC under grant number TG-DMR090107T.

REFERENCES

[1] Boek, E.S., Coveney, R., Skipper, N.T.: Molecular modeling of clay hydration: A study of hysteresis loops in the swelling curves of sodium montmorillonite. Langmuir 11, 4629–4631 (1995)
[2] Vanorio, T., Prasad, M., Nur, A.: Elastic properties of dry clay mineral aggregates, suspensions and sandstones. Geophys. J. Int. 155, 319–326 (2003)
[3] Chen, B., Evans, J.: Elastic moduli of clay platelets. Scripta Mater. 54, 1581–1585 (2006)
[4] Sainz-Diaz, C.I., Hernandez-Laguna, A., Dove, M.T.: Modelling of dioctahedral 2:1 phyllosilicates by means of transferable empirical potentials. Phys. Chem. Miner. 28, 130–141 (2001)
[5] de Leeuw, N.H., Parker, S.C.: Molecular-dynamics simulation of MgO surfaces in liquid water using a shell-model potential for water. Phys. ReV. B 58, 13901–13908 (1998)
[6] Higgins, F.M., de Leeuw, N.H., Parker, S.C.: Modelling the effect of water on cation exchange in zeolite A. J. Mater. Chem. 12, 124–131 (2002)
[7] Cygan, R.T., Liang, J.-J., Kalinichev, A.G.: Molecular models of hydroxide, oxyhydroxide, and clay phases and the development of a general forcefield. J. Phys. Chem. B. 108, 1255–1266 (2004)
[8] Gale, J.D., Rohl, A.L.: The General Utility Lattice Program. Mol. Simul. 29, 291–341 (2003), `https://www.ivec.org/gulp`
[9] Refson, K., Park, S.H., Sposito, G.: Ab initio computational crystallography of 2: 1 clay minerals: 1. Pyrophyllite-1Tc. J. Phys. Chem. B 107, 13376–13383 (2003)
[10] Fu, M.H., Zhang, Z.Z., Low, P.F.: Changes in the properties of a montmoriUonite-water system during the adsorption and desorption of water: Hysteresis. Clays & Clay Minerals 38, 485–492 (1990)
[11] Pawley, A.R., Clark, S.M., Chinnery, N.J.: Equation of state measurements of chlorite, pyrophylite, and talc. American Mineralogist 87, 1172–1182 (2002)
[12] McNeil, L.E., Grimsditch, M.: Elastic moduli of muscovite mica. J. Phys.: Condens. Matter. 5, 1681–1690 (1993)
[13] Prasad, M., Kopycinska, M., Rabe, U., Arnold, W.: Measurement of Young's modulus of clay minerals using atomic force acoustic microscopy. Geophys. Res. Lett. 29, 1172 (2002)

EMERGENT CHEMO-HYDRO-MECHANICAL PHENOMENA IN CARBON GEOLOGICAL STORAGE

D. Nicolas Espinoza and J. Carlos Santamarina

School of Civil and Environmental Engineering
Georgia Institute of Technology
790 Atlantic Dr., Atlanta, GA 30332
webpage: http://pmrl.ce.gatech.edu

Summary. The combustion of fossil fuels produces carbon dioxide (CO_2). Carbon capture and geological storage (CCS) has been proposed to reduce the emission of greenhouse gases to the atmosphere. The properties of CO_2 depend on pressure and temperature: it can be found as a gas, liquid, or supercritical, and it reacts with water to produce carbonic acid lowering the water pH and can form a solid hydrate mass. These characteristics give rise to complex chemo-hydro-thermo-mechanical coupled processes and emergent phenomena that can condition the long-term geological storage of CO_2. Processes include wettability, leakage, mineral dissolution, and CH_4-CO_2 replacement in hydrate-bearing sediments.

Keywords: CO_2, coupled phenomena, gas breakthrough, leakage.

1 INTRODUCTION

The geological disposal of carbon dioxide (CO_2) is an attempt towards a more sustainable use of fossil fuels. The target formations for disposal include saline formations, oil/gas reservoirs, coal seams, and hydrate-bearing sediments. Pressure and temperature conditions range from cold and relatively shallow permafrost and marine sediments

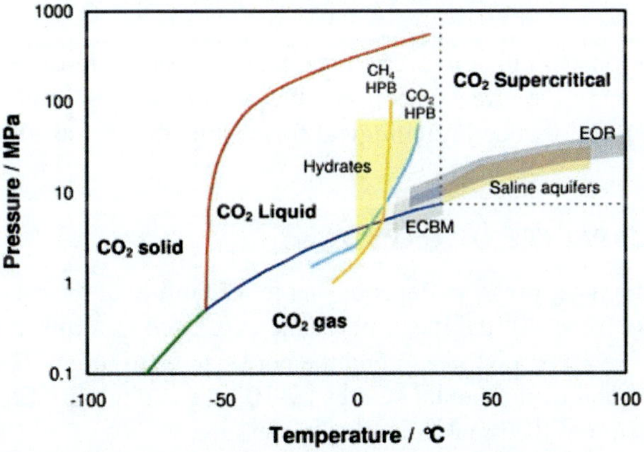

Fig. 1. Pressure–temperature conditions in CO_2 geological storage projects. This figure is based on ongoing and planned projects worldwide [3, 4].

R.I. Borja (Ed.): Multiscale and Multiphysics Processes in Geomechanics, SSGG, pp. 109–112.
springerlink.com

(e.g., Alaska North slope: 7MPa, 278K), and warm coal seams (e.g., Alabama Black Warrior basin: 7MPa, 296K) to deep hot rocks onshore (e.g., Weyburn oil field: 14MPa, 323K). Depending on pressure and temperature, CO_2 can be stored as gas, liquid, supercritical phase, or solid hydrate mass (Figure 1).

The physical properties of CO_2 depend strongly on temperature and pressure, including density, bulk modulus, viscosity, and its interaction with water (e.g., solubility and interfacial tension) [1, 2]. The reactivity of CO_2 with water, organic compounds, and minerals gives rise to phenomena, such as water acidification, coal swelling and decreased viscosity of hydrocarbons, and mineral dissolution and changes in sediment mechanical and hydraulic conditions. These phenomena couple at the pore and particle scales and lead to the emergence of complex phenomena at the reservoir scale during and after CO_2 injection. Selected phenomena and implications are briefly discussed in the next section.

2 INTERFACIAL PHENOMENA: WETTABILITY

Interfacial tension and contact angle in CO_2–water–mineral systems upscale through the sediment porous network to define multiphase flow characteristics such as residual saturation, relative permeability, imbibition patterns, gas-entry pressure, and suction [5, 6]. Changes in suction can cause a pronounced impact in the reservoir state of stress [7].

Our results show that CO_2–water interfacial tension decreases significantly from $\sim$72 to 25mN/m as pressure increases from 0.1MPa to 6.4MPa (T =298K) in agreement with previous studies [8–10]. Contact angle varies with CO_2 pressure in response to changes in CO_2–water interfacial tension: it increases on non-wetting surfaces such as PTFE and oil-wet quartz and slightly decreases in water-wet quartz and calcite surfaces. Both interfacial tension σ and contact angle θ combine to define the capillary pressure p_c^*, i.e., the pressure difference between invading and host fluids; hence, the invading fluid can penetrate a medium with characteristic pore diameter d^*,

$$p_c^* = \frac{4\sigma \cos\theta}{d^*} \quad (1)$$

Combining these observations, we can anticipate that pressure-dependent interfacial tension and contact angle affect injection patterns and breakthrough mechanisms, in other words, the performance of geological formations that act as either reservoirs or seals.

3 SEALING CAPACITY OF CAPROCKS

The breakthrough gas pressure depends on the wettability of the rock in the presence of water and CO_2 ($\sigma \cos\theta$), the mean pore size expressed in terms of specific surface S_s and void ratio $e = f(e_{1\text{kPa},C_c,p'})$, and the pore size distribution. Based on previous data compilations and experimental studies [8–10], we can modify Eq. 1 to obtain the following expression of the breakthrough gas pressure:

$$p_c^* = \psi \frac{S_s \rho \sigma \cos\theta}{e_{1\text{kPa}} - C_c \log \frac{p'}{1\text{kPa}}} \quad (2)$$

Our current experimental studies explore the applicability of this expression for overconsolidated clay plugs in long-term storage conditions.

4 MINERAL DISSOLUTION

CO_2-acidified water reacts with the solid mineral skeleton and causes mineral dissolution. In particular, carbonate rocks react with carbonic acid at a much faster rate than siliceous minerals [11]. Water acidification and calcium carbonate dissolution reactions are as follows:

$$H_2O(l) + CO_2(aq) \rightleftharpoons H_2CO_3(aq) \text{ carbonic acid}$$
$$CaCO_3(s) + H_2O + CO_2(g) \rightleftharpoons Ca^{2+} + 2\,HCO_3^-$$

Mineral dissolution at the particle scale has two very important macro-scale implications: (1) changes in the matrix permeability with the possibility of fluid flow localization and piping [12], and (2) decreased horizontal effective stress reaching internal failure under zero lateral strain conditions, possibly causing shear strain localization [13]. These two coupled phenomena, hydro-chemo and chemo-mechanical, and the emergence of localizations threaten the long-term geological storage of CO_2.

5 CH_4-CO_2 REPLACEMENT IN HYDRATE-BEARING SEDIMENTS

There are very large reserves of methane (CH_4) in hydrate form, filling the pore space in sediments. Thermodynamic conditions and previous experimental evidence confirm that CH_4 may be replaced by CO_2 in clathrate hydrates [14, 15]. A detailed analysis of the multiple coexisting processes and new experimental results allow us to anticipate potential geomechanical implications during CH_4-CO_2 replacement in hydrate-bearing sediments [16]. While the solid nature of bulk hydrate remains contributing stability to the sediment skeleton, other processes may determine the implications and effectiveness of CH_4-CO_2 replacement, including pronounced fluid volume change, "water drying", additional hydrate mass formation and pore plugging, and invasion by viscous fingering among others.

REFERENCES

[1] Span, R., Wagner, W.: A new equation of state for carbon dioxide covering the fluid region from the triple-point temperature to 1100 K at pressures up to 800 MPa. Journal of Physical and Chemical Reference Data 25(6), 1509–1596 (1996)

[2] Fenghour, A., et al.: The viscosity of carbon dioxide. Journal of Physical and Chemical Reference Data 27(1), 31–44 (1998)

[3] Metz, B., et al. (eds.): Prepared by Working Group III of the Intergovernmental Panel on Climate Change, IPCC Special Report on Carbon Dioxide Capture and Storage IPCC 2005, p. 442. Cambridge University Press, Cambridge (2005)

[4] DOE-NETL, Carbon sequestration atlas of the United States of America and Canada, 2nd edn. (2008)

[5] Plug, W.J., Bruining, J.: Capillary pressure for the sand-CO2-water system under various pressure conditions. Application to CO2 sequestration. Advances in Water Resources 30, 2339–2353 (2007)

[6] Chalbaud, C., et al.: Interfacial tension measurements and wettability evaluation for geological CO2 storage. Advances in Water Resources 32(1), 98–109 (2009)

[7] Delage, P., et al.: Subsidence and capillary effects in chalks, paper presented at Dans EU-ROCK 1996. In: Prediction and Performance in Rock Mechanics and Rock Engineering - ISRM International symposium, Torino, France (1996)
[8] Hildenbrand, A., et al.: Gas breakthrough experiments on pelitic rocks: comparative study with N-2, CO2 and CH4. Geofluids 4(1), 61–80 (2004)
[9] Li, S., et al.: Gas breakthrough pressure for hydrocarbon reservoir seal rocks: implications for the security of long-term CO2 storage in the Weyburn field. Geofluids 5(4), 326–334 (2005)
[10] Espinoza, D.N., Santamarina, J.C.: Water-CO2-mineral systems: interfacial tension, contact angle and diffusion Implications to CO2 geological storage. Water Resources Research 46, W07537 (2010), doi:10.1029/2009WR008634
[11] Emberley, S., et al.: Geochemical monitoring of fluid-rock interaction and CO2 storage at the Weyburn CO2-injection enhanced oil recovery site, Saskatchewan, Canada. Energy 29(9-10), 1393–1401 (2004)
[12] Gunter, W.D., et al.: Aquifer disposal of acid gases: modelling of water-rock reactions for trapping of acid wastes. Applied Geochemistry 15(8), 1085–1095 (2000)
[13] Shin, H., et al.: Contraction-driven shear failure in compacting uncemented sediments. Geology 36(12), 931–934 (2008)
[14] Stevens, C.J., et al.: Experimental hydrate formation and gas production scenarios based on CO2 sequestration. In: Paper Presented at Proceedings of the 6th International Conference on Gas Hydratess, Vancouver, British Columbia, Canada (2008)
[15] McGrail, B.P., et al.: Using carbon dioxide to enhance recovery of methane from gas hydrate reservoirs: final summary report. PNNL-17035, Pacific Northwest National Laboratory operated by Battelle Memorial Institute for the U.S. Department of Energy, Oak Ridge, TN (2007)
[16] Jung, J.W., Espinoza, D.N., Santamarina, J.C.: Hydrate Bearing Sediments: Properties and Phenomena relevant to CH4-CO2 Replacement. Journal of Geophysical Research-Solid Earth (in press), doi:10.1029/2009JB000812RR

ISOGEOMETRIC FAILURE ANALYSIS

René de Borst[1], Thomas J.R. Hughes[2],
Michael A. Scott[2], and Clemens V. Verhoosel[1,2]

[1] Department of Mechanical Engineering
Eindhoven University of Technology
P.O. Box 513, 5600 MB Eindhoven, The Netherlands
e-mail: r.d.borst,c.v.verhoosel@tue.nl, web page: http://w3.wtb.tue.nl
[2] Institute for Computational Engineering and Sciences
University of Texas at Austin
Austin, Texas 78712, U.S.A.
e-mail: {hughes,mscott}@ices.utexas.edu, web page: http://www.ices.utexas.edu

Summary. The isogeometric method is a versatile tool for failure analysis. On one hand, the high-order continuity enables a natural incorporation of continuum constitutive relations that incorporate higher-order strain gradients, as in gradient plasticity or damage. On the other hand, the possibility of enhancing a T-spline basis with discontinuities by means of knot insertion makes isogeometric finite elements a suitable candidate for modeling discrete cracks. Both possibilities are described and will be illustrated by examples.

Keywords: cohesive zones, fracture, gradient damage, isogeometric analysis, NURBS, T-splines.

1 INTRODUCTION

Understanding and predicting failure is of crucial importance for improving the design of engineering structures. Two distinct approaches have found their way into the literature: smeared and discrete models. Traditional smeared crack models are limited by the fact that a smeared decohesion relation induces a loss of well-posedness of the rate boundary value problem at a generic stage in the loading process. To regularize this ill-posedness, constitutive relations have been enhanced with higher-order spatial gradients. Traditional finite elements have difficulties to accommodate the ensuing higher-order continuity. Isogeometric analysis, which employs spline functions such as Non-Uniform Rational B-Splines (NURBS), however, possesses the required higher-order continuity naturally, and is therefore a perfect tool for dealing with such enhanced constitutive models. From the perspective of element technology, the challenge of capturing discrete fracture models lies in flexibly capturing the internal free boundaries. This is particularly so when propagating discontinuities are to be simulated. Among the available finite element technologies for capturing propagating discontinuities are interface elements (e.g. [1]), embedded discontinuities (e.g. [2]), and the partition-of-unity method (e.g. [3–5]).

In this contribution, isogeometric analysis, as introduced by Hughes et al. [6, 7], is used to analyze gradient-enhanced continuum damage models and to simulate discrete cracks that employ a cohesive zone concept. One possibility is to use isogeometric finite elements in combination with the partition of unity method. In such a case, the

R.I. Borja (Ed.): Multiscale and Multiphysics Processes in Geomechanics, SSGG, pp. 113–116.
springerlink.com

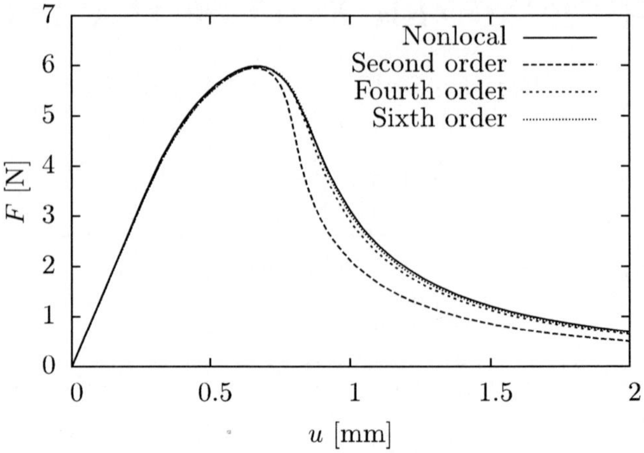

Fig. 1. Force–displacement diagrams for the three-point bending simulation using the nonlocal formulation and the second-, fourth-, and sixth-order formulations.

discontinuities would be embedded in the solution space by means of Heaviside functions. Although such an approach would benefit from both advantages of the isogeometric approach, isogeometric finite elements offer the possibility to directly insert discontinuities in the solutions space. The conceptual idea is that in the isogeometric approach the inter-element continuity can be decreased by means of knot insertion. Knots should not be confused with finite element nodes, although the proposed concept of treating discontinuities is similar to that in interface elements.

2 NUMERICAL EXAMPLES

2.1 Gradient Damage: Three-Point Bending Beam

As an illustration of the application of isogeometric analysis to gradient-damage models, we consider a three-point bending experiment. The specimen is loaded by a distributed load over the central area of the beam. The force–displacement curves shown in Figure 1 are obtained using a nonlocal integral formulation and the second-, fourth-, and sixth-order gradient damage formulations. For the second- and fourth-order gradient formulations, 4704 quadratic NURBS elements are used. For the sixth-order formulation, 4704 cubic NURBS elements have been used. The nonlocal result is obtained using 1200 quadratic NURBS elements. Upon increasing the order of the formulation, the force–displacement curve converges to that of the nonlocal integral model, as expected.

2.2 Cohesive Zone Model: Single-Edge Notched Beam

The next simulation is that of the single-edge notched beam. The anti-symmetric loading conditions applied to the specimen create a curved crack that nucleates from the initial notch in the upper edge. Following the first successful numerical simulation [9],

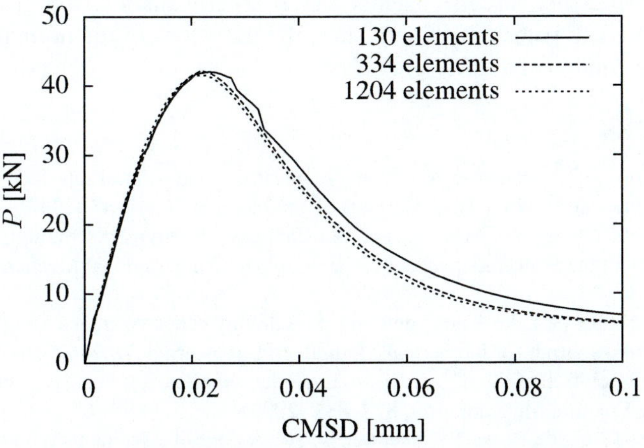

Fig. 2. Response curves for the SEN-beam simulations.

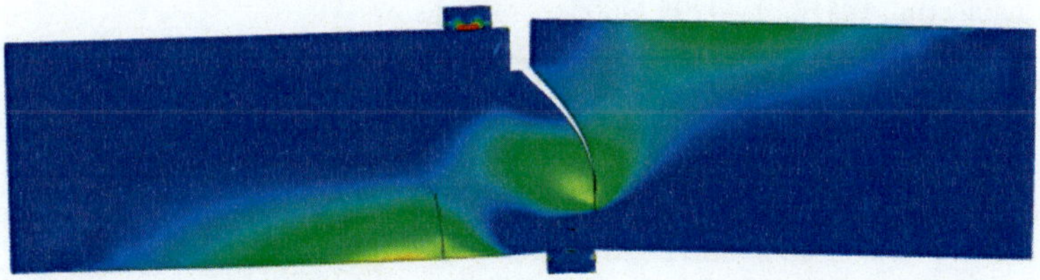

Fig. 3. Contour plot showing the $\sigma_{x_1x_1}$ Cauchy stress in the SEN-beam at CMSD=0.033 mm using the finest discretization.

several numerical techniques have been used, such as the partition of unity method [3] and gradient-enhanced continuum damage [10].

The beam geometry and the deformation are described using a cubic T-spline mesh [8]. The coarsest mesh which we have considered consists of 130 elements (402 DOFs). Two uniform mesh refinements have been performed, with 334 elements (868 DOFs) and 1204 elements (2734 DOFs), respectively. The equilibrium path is traced using the crack mouth sliding displacement (CMSD), which is defined as the difference in vertical displacement between the notch tips, as the path parameter [9]. The response of the SEN beam is shown in Figure 2. The response is measured in terms of the exerted force P versus the CMSD. It is observed that the result using the intermediate mesh practically coincides with that of the fine mesh. It is therefore concluded that an accurate solution could be obtained even with only 334 elements.

In Figure 3, a contour plot of the cracked SEN-beam is shown. The dominant crack nucleates at the bottom right corner of the initial notch at an angle of 37 degrees with the x_2-axis. Upon extension, the crack gradually turns to an angle of zero degree with the vertical axis, which is in good agreement with experimental observations. From the contour plot, it can also be observed that a secondary crack nucleates at the bottom edge of the specimen, which has also been observed experimentally. Note from the contour plot that both crack paths are smooth since the directions of the normal vectors from one segment to another have been matched.

REFERENCES

[1] Schellekens, J.C.J., de Borst, R.: On the numerical integration of interface elements. International Journal for Numerical Methods in Engineering 36, 43–43 (1993)
[2] Simo, J.C., Oliver, J., Armero, F.: An analysis of strong discontinuities induced by strain-softening in rate-independent inelastic solids. Computational Mechanics 12, 277–296 (1993)
[3] Wells, G.N., Sluys, L.J.: A new method for modelling cohesive cracks using finite elements. International Journal for Numerical Methods in Engineering 50, 2667–2682 (2001)
[4] Moës, N., Belytschko, T.: Extended finite element method for cohesive crack growth. Engineering Fracture Mechanics 69, 813–833 (2002)
[5] Remmers, J.J.C., de Borst, R., Needleman, A.: A cohesive segments method for the simulation of crack growth. Computational mechanics 31, 69–77 (2003)
[6] Hughes, T.J.R., Cottrell, J.A., Bazilevs, Y.: Isogeometric analysis: CAD, finite elements, NURBS, exact geometry and mesh refinement. Computer Methods in Applied Mechanics and Engineering 194, 4135–4195 (2005)
[7] Cottrell, J.A., Hughes, T.J.R., Bazilevs, Y.: Isogeometric analysis: Towards integration of CAD and FEA. Wiley, Chichester (2009)
[8] Sederberg, T.W., Zheng, J., Bakenov, A., Nasri, A.: T-splines and T-NURCCs. ACM Transactions on Graphics 22, 477–484 (2003)
[9] de Borst, R.: Computation of post-bifurcation and post-failure behavior of strain-softening solids. Computers & Structures 25, 211–224 (1987)
[10] Geers, M.G.D., de Borst, R., Peerlings, R.H.J.: Damage and crack modeling in single-edge and double-edge notched concrete beams. Engineering Fracture Mechanics 65, 247–261 (2000)

MICRO-CRACK INFORMED MULTI-SCALE DAMAGE MODEL: THEORY AND COMPUTATION

Xiaodan Ren[1,2], Jiun-Shyan Chen[2], and Jie Li[1]

[1] School of Civil Engineering
Tongji University
1239 Siping Road, Shanghai 200092, China
e-mail: lijie@tongji.edu.cn
[2] Department of Civil and Environmental Engineering
University of California
P.O. Box 159310, Los Angeles, CA 90095-1593, USA
e-mail: jschen@seas.ucla.edu, web page: http://www.cee.ucla.edu/faculty/jschen.htm

Summary. A class of multi-scale damage models for the softening of the brittle solids induced by micro-cracks is proposed, in which the damage evolution is the direct consequence of the micro-crack propagation. The homogenization of stress and strain fields in the cracked unit cell yields the degradation tensor, and the equivalence between the averaged strain energy of the unit cell and the strain energy density of the homogenized material is demonstrated. This energy equivalence relationship is consistent with that obtained from asymptotic-based homogenization, and it serves as an energy bridging vehicle between damaged continua and cracked microstructure. Damage evolution equations are obtained by this energy-bridging method. The size effect of the homogenized damage law is also characterized through the unit cell analysis, and the results are used to eliminate the mesh dependencies in damage-induced strain localization problems.

Keywords: damage model, Helmholtz free energy, homogenization, micro-crack.

1 HOMOGENIZED AND AVERAGED KINEMATICS AND KINETICS

Let a heterogeneous solid with domain Ω and boundary Γ containing a distribution of micro-cracks be considered. For a given material point in the macroscopic solid, it corresponds to a unit cell microstructure with domain, Ω_y and $\partial\Omega_y$, which contains micro-crack surface Γ_c as shown in Fig.1. The tractions and displacements prescribed on the boundary of the microscopic unit cell $\partial\Omega_y$ are related to the homogenized stress and strain in the continua as

$$\overline{\boldsymbol{\sigma}} = \frac{1}{V_y} \oint_{\partial\Omega_y} (\boldsymbol{t}^{\varepsilon} \otimes \boldsymbol{x}^{\varepsilon}) ds \tag{1}$$

$$\overline{\boldsymbol{e}} = \frac{1}{2V_y} \oint_{\partial\Omega_y} (\boldsymbol{u}^{\varepsilon} \otimes \boldsymbol{n} + \boldsymbol{n} \otimes \boldsymbol{u}^{\varepsilon}) ds \tag{2}$$

R.I. Borja (Ed.): Multiscale and Multiphysics Processes in Geomechanics, SSGG, pp. 117–120.
springerlink.com

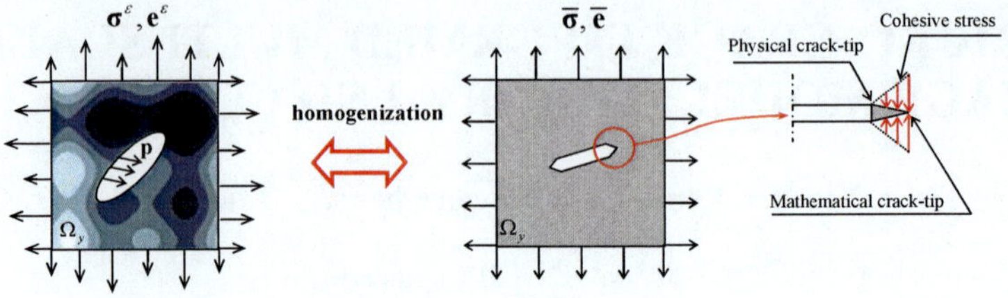

Fig. 1. Homogenization of unit cell with fluctuating fields.

where $V_y = \int_{\Omega_y} d\Omega$ is the volume of the unit cell. Alternatively, the averaged stress and strain in the unit cell are defined as

$$\langle \boldsymbol{\sigma} \rangle = \frac{1}{V_y} \int_{\Omega_y} \boldsymbol{\sigma}^\varepsilon d\Omega \tag{3}$$

$$\langle \boldsymbol{e} \rangle = \frac{1}{V_y} \int_{\Omega_y} \boldsymbol{e}^\varepsilon d\Omega \tag{4}$$

By substituting $\boldsymbol{\nabla} \cdot (\boldsymbol{\sigma}^\varepsilon \otimes \boldsymbol{x}) = \boldsymbol{\nabla} \cdot \boldsymbol{\sigma}^\varepsilon \otimes \boldsymbol{x} + \boldsymbol{\sigma}^\varepsilon \cdot (\boldsymbol{\nabla} \otimes \boldsymbol{x}) = \boldsymbol{\sigma}^\varepsilon$ in averaged stress definition (3) and consider, the equilibrium of the cohesive stresses on the crack surface [1], one gets

$$\overline{\boldsymbol{\sigma}} = \langle \boldsymbol{\sigma} \rangle \tag{5}$$

$$\overline{\boldsymbol{e}} = \langle \boldsymbol{e} \rangle + \frac{1}{2V_y} \oint_{\Gamma_c} (\boldsymbol{u}^\varepsilon \otimes \boldsymbol{n} + \boldsymbol{n} \otimes \boldsymbol{u}^\varepsilon) ds \tag{6}$$

2 ENERGY BRIDGING

To obtain the relationship between the micro-cracks induced material degradation and the continuum damage mechanics, the Helmholtz free energy (HFE) is employed:

$$\overline{\Psi} = \frac{1}{2} \overline{\boldsymbol{\sigma}} : \overline{\boldsymbol{e}} \tag{7}$$

According to the second principle of thermodynamics, we have

$$\overline{\boldsymbol{\sigma}} = \frac{\partial \overline{\Psi}}{\partial \overline{\boldsymbol{e}}}, \boldsymbol{Y} = -\frac{\partial \overline{\Psi}}{\partial \boldsymbol{D}}, \dot{\boldsymbol{D}} = \dot{\lambda} \frac{\partial \overline{\Psi}}{\partial \boldsymbol{Y}} \tag{8}$$

where $\boldsymbol{D}$ is the damage tensor and $\boldsymbol{Y}$ is the damage energy release rate (DERR) representing a driving force of damage evolution. In the conventional continuum damage mechanics, HFE is determined experimentally. In the present approach, the HFE is obtained through the homogenization of cracked microstructure. The microscopic free energy is the elastic strain energy defined as

$$\Psi^\varepsilon = \frac{1}{2} \boldsymbol{\sigma}^\varepsilon : \boldsymbol{e}^\varepsilon \tag{9}$$

where σ^{ε} and e^{ε} are the microscopic stress and strain, respectively. Integrating microscopic HFE in the unit cell yields

$$\int_{\Omega_y} \Psi^{\varepsilon} d\Omega = \frac{1}{2} \oint_{\partial\Omega_y} \boldsymbol{u}^{\varepsilon} \cdot \boldsymbol{t}^{\varepsilon} ds - \frac{1}{2} \oint_{\partial\Gamma_c} \boldsymbol{u}^{\varepsilon} \cdot \boldsymbol{p} ds \tag{10}$$

By considering the prescribed boundary displacements on the unit cell from the macroscopic strain by $\boldsymbol{u}^{\varepsilon} = \overline{\boldsymbol{e}} \cdot \boldsymbol{x}$ on $\partial\Omega_y$, we have

$$\frac{1}{2} \oint_{\partial\Omega_y} \boldsymbol{u}^{\varepsilon} \cdot \boldsymbol{t}^{\varepsilon} ds = \frac{V_y}{2} \overline{\boldsymbol{\sigma}} : \overline{\boldsymbol{e}} = V_y \overline{\Psi} \tag{11}$$

Combining Eqs. (10) and (11), we have

$$\overline{\Psi} = \frac{1}{V_y} \left(\int_{\Omega_y} \Psi^{\varepsilon} d\Omega + \frac{1}{2} \oint_{\partial\Gamma_c} \boldsymbol{u}^{\varepsilon} \cdot \boldsymbol{p} ds \right) \tag{12}$$

The right hand side of (12) is the averaged energy density of the unit cell, while the left-hand side of (12) is the energy density of the homogenized material [2]. We use the Helmholtz free energy relationship between the homogenized continuum and cracked microstructure to derive the damage parameters for several commonly used damage models.

3 UNIT CELL ANALYSIS

The essential step in obtaining the damage evolution function in the proposed method is the unit cell analysis. In this study, the unit cell is made of an isotropic linear elastic material with a center crack subjected to pre-defined boundary conditions as shown in Fig. 1. The boundary conditions are defined based on the type of damage model to be employed. In this study, we consider an enriched reproducing kernel particle method (RKPM) for the unit cell analysis [3]. J-integral is used in the crack propagation criteria. The predicted unit cell solution is then used in the homogenization procedures as described in the previous section for characterization of the damage parameters evolution.

4 NUMERICAL EXAMPLES

A notched beam under three-point bending as shown in Fig. 2 is to be modeled by the proposed methods. The material properties of elastic modulus and Poisson's ratio are 30 GPa and 0.2, respectively. The tensile strength of the material is $f_u = 3.33$MPa, and the fracture energy of cohesive crack is adopted as $G_I = 124$N/m.

The homogenized stress-strain curves are strongly affected by the size of the unit cell as shown in Fig. 3a. The size dependency of the nominal strength agrees well with the size effect law proposed by Bazant et al (1982) [4] as shown in Fig. 3a. These similar characteristics are due to the influence of the physical length scale embedded in the crack opening law, to the overall deformation of material under consideration. The characterized damage law for isotropic one-parameter damage model is employed in the structural level analysis. To study the mesh sensitivities of the multiscale analysis, coarse, medium,

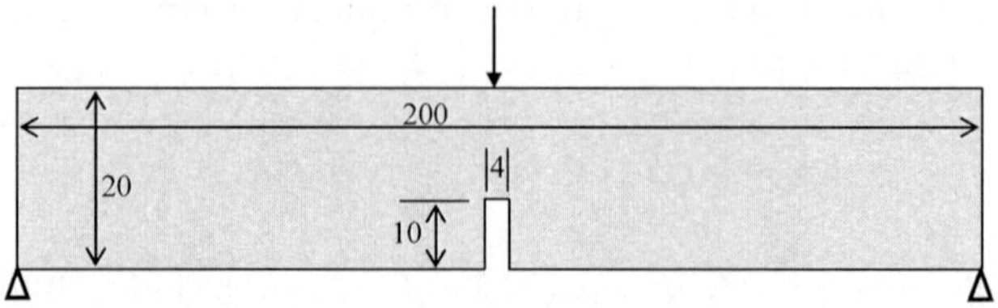

Fig. 2. Notched beam subjected to three-point bending (unit: cm).

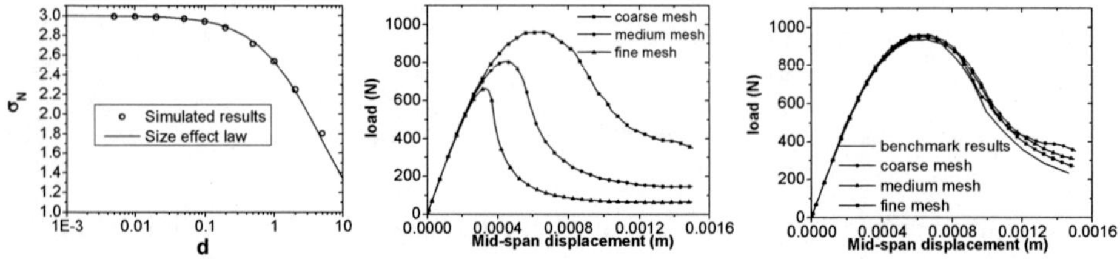

Fig. 3. (a) numerical and physical size effects, (b) mesh-dependent results using conventional continuum damage model, and (c) mesh independent results using the proposed method.

and fine meshes are employed for the notched beam. In the conventional damage models, the damage evolution curves are directly used in the structural analysis without consideration of the micro structures and mesh sizes. Figure 3b shows a strong mesh dependency induced by this standard procedure. With the proposed method, the micro-crack induced damage evolution curves for three unit cell dimensions corresponding to the three macroscopic element dimensions are first calculated. By introducing damage evolution curves consistent with the corresponding structural meshes, the mesh independent results are obtained as shown in Fig. 3c.

REFERENCES

[1] Nemat-Nasser, S., Hori, M.: Micromechanics: overall properties of heterogeneous materials, 2nd edn. Elsevier Science B.V., Amsterdam (1999)

[2] Hill, R.: Elastic properties of reinforced solids: some theoretical principles. J. Mech. Phys. Solids 11, 357–372 (1963)

[3] Chen, J.S., Pan, C.H., Tu, C.T., Liu, W.K.: Reproducing kernel particle methods for large deformation analysis of non-linear structures. Computer Methods in Applied Mechanics and Engineering 139, 195–227 (1996)

[4] Bazant, Z.P.: Size effect in blunt fracture: concrete, rock, metal. Journal of Engineering Mechanics, ASCE 110(4), 518–535 (1984)

CIRCUMVENTING THE VOLUME AND CONTACT CONSTRAINTS IN MULTIPHYSICS FINITE ELEMENT SIMULATIONS

Ronaldo I. Borja[1], Joshua A. White[2], and Fushen Liu[3]

[1] Department of Civil and Environmental Engineering
Stanford University
Stanford, CA94305, USA
e-mail: borja@stanford.edu
[2] Computational Geosciences Group
Lawrence Livermore National Laboratory
P.O. Box 808, L-231, Livermore, CA, 94551, USA
e-mail: jawhite@llnl.gov
[3] Department of Civil and Environmental Engineering
Stanford University
Stanford, CA94305, USA
e-mail: fsliu@stanford.edu

Summary. We consider two multiphysics problems in geomechanics with constraints. The first involves coupled solid deformation–fluid diffusion subject to a volume constraint in the limit of no-flow. The second is the classic contact problem subject to a no-interprenetration constraint. Both problems are known to generate unphysical oscillation in the pressure field when some combinations of displacement and pressure finite element interpolations are employed. We trace the problem to the non-satisfaction of the LBB condition and utilize a stabilization technique based on polynomial pressure projection to circumvent the unpreferable pressure oscillation.

Keywords: contact problems, coupled processes, extended finite element, stabilized finite element.

1 INTRODUCTION

There exists a large body of literature addressing the computational aspects of multiphysics problems subject to a volume constraint, as well as to contact problems subject to the restriction of no interpenetration between contacting faces. In both problems, either the continuum pressure or the normal contact stress exhibits spurious oscillation. This oscillation does not go away with mesh refinement, and in some cases, it even gets worse as the mesh is refined. We trace this problem to the non-satisfaction of the LBB condition [2] associated with equal-order interpolation of displacement and pressure degrees of freedom for the case of continuum multiphysics problem, or with equal-order interpolation of slip and normal component of traction for the case of the contact problem. In this article, we employ a stabilized finite element formulation based on the polynomial pressure projection (PPP) technique, which was used successfully for Stokes equation [1]. For the frictional contact problem, the PPP approach is applied to

R.I. Borja (Ed.): Multiscale and Multiphysics Processes in Geomechanics, SSGG, pp. 121–124.
springerlink.com

the normal contact pressure in the framework of the extended finite element method. We use low-order finite elements for both problems: linear pair for displacement and pressure degrees of freedom for the multiphysics continuum problem, and linear triangular elements (tetrahedral elements for 3D) for slip and normal pressure degrees of freedom for the contact problem.

2 STABILITY CONDITION

For Stokes equation, it is known that the discrete velocity and pressure spaces, $\mathcal{U}^h$ and $\mathcal{P}^h$, respectively, must be chosen to satisfy the discrete LBB condition:

$$\sup_{\boldsymbol{v}^h \in \mathcal{U}^h} \frac{\int_\Omega \psi^h \nabla \cdot \boldsymbol{v}^h \, \partial\Omega}{\|\boldsymbol{v}^h\|_1} \geq C\|\psi^h\|_0 \qquad \forall \psi^h \in \mathcal{P}^h, \tag{1}$$

with $C > 0$ being independent of h. Unfortunately, many linear-pressure/linear-velocity interpolations do not satisfy this condition and lead to unstable approximations. However, Bochev and co-workers (see [1] and references therein) demonstrated that this linear pair does satisfy the weaker condition:

$$\sup_{\boldsymbol{v}^h \in \mathcal{U}^h} \frac{\int_\Omega \psi^h \nabla \cdot \boldsymbol{v}^h \, \partial\Omega}{\|\boldsymbol{v}^h\|_1} \geq C_1\|\psi^h\|_0 - C_2\|\psi^h - \Pi\psi^h\|_0 \qquad \forall\, \psi^h \in \mathcal{P}^h, \tag{2}$$

where $\Pi : L_2(\Omega) \to R_0$, and with $C_1 > 0$ and $C_2 > 0$ being independent of h. The term $C_2\|q^h - \Pi q^h\|_0$ quantifies the inherent deficiency in the linear pair. Motivated by this development, White and Borja [4] successfully adopted a similar stabilization for coupled solid deformation–fluid diffusion problem, with the "velocity" now interpreted as pertaining to the motion of the solid matrix and the pore fluid pressure taking the role of the generic "pressure" term. More recently, and in a similar vein, Liu and Borja [3] reformulated the contact problem as a Lagrange multipliers problem in the framework of the extended finite element method. They also adopted a similar stabilization scheme successfully, with the slip degree of freedom taking the role of the "velocity" term and the contact pressure playing the role of the generic "pressure" term.

3 NUMERIC EXAMPLES

The first example demonstrates the efficacy of the stabilization scheme for handling the volume constraint imposed by the lack of fluid flow in a coupled solid deformation–fluid flow simulation. We consider a three-dimensional square footing pressed vertically into a fully saturated ground. For the same number of mixed finite elements, a linear-pair displacement/pressure interpolation (Q1Q1) yields a total of 19,652 degrees of freedom, whereas a quadratic displacement/linear pressure interpolation (Q2Q1) results in a total of 112,724 degrees of freedom. The Q1Q1 pair is known to be unstable, and Figure 1(a) corroborates this with the unpreferable pressure oscillation resulting from the unstabilized formulation. The Q2Q1 pair is known to be stable, and consequently Figure 1(b) shows a smooth variation of pressure within the problem domain. After stabilizing the Q1Q1 pair, the pressure oscillation disappears, as shown in Figure 1(c).

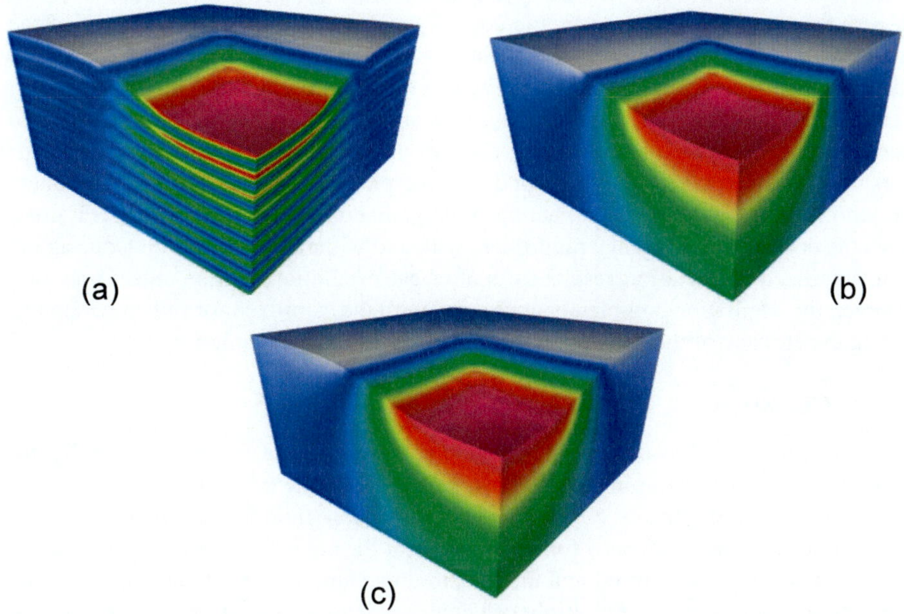

Fig. 1. Square footing pressed vertically into a fully saturated ground: (a) unstabilized linear-pair hexahedral elements; (b) stable quadratic displacement/linear pressure hexahedral elements; and (c) stabilized linear-pair hexahedral elements, based on [5]

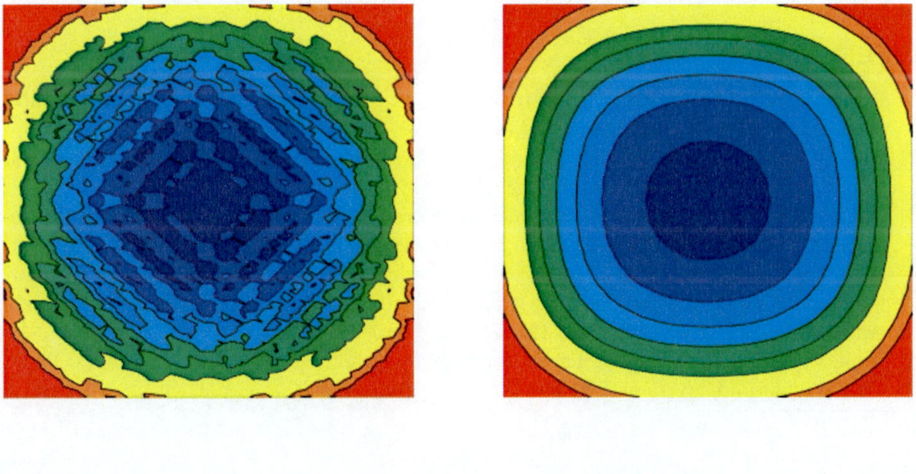

Fig. 2. Unstabilized (left) and stabilized (right) contact pressures on a pair of smooth cubes pressed against each other, based on [3].

The second example demonstrates the efficacy of the stabilization scheme for dealing with the constraint imposed by the contact condition within the framework of the extended finite element method. We consider an elastic cube clamped at both its top and bottom ends and pressed in the vertical direction. The cube is discretized with three-dimensional tetrahedral elements, and a smooth horizontal plane is embedded midway between the top and bottom surfaces of the cube. We use the Lagrange multipliers method to impose the contact condition and plot the resulting normal component of traction (Lagrange multipliers) acting on the embedded horizontal plane. Figure 2 shows the contours of Lagrange multipliers with and without stabilization. Once again, the unstabilized formulation results in unphysical oscillation in the contact pressure. However, the oscillation completely goes away with the stabilized formulation, demonstrating the efficacy of the stabilization technique for contact problems.

4 CONCLUSION

The polynomial pressure projection technique, which was used successfully for Stokes flow, is shown to minimize if not completely eliminate spurious oscillation of the pore pressure in multiphysics coupled solid deformation–fluid flow problem. It is also shown to exhibit similar efficacy for stabilization of the contact stress for both stick and slip conditions, with structured and unstructured meshes, with equal and unequal line segments, and with Lagrange multipliers, penalty, and augmented Lagrangian methods. Additional test problems reported in [3, 4] included 2D plane stress, 2D plane strain, and full 3D multiphysics and contact problems.

ACKNOWLEDGMENTS

This study is supported by the Office of Basic Energy Sciences, the U.S. Department of Energy, under Grant No. DE-FG02-03ER15454, and by the U.S. National Science Foundation, under Contract Numbers CMMI-0824440 and CMMI-0936421 to Stanford University.

REFERENCES

[1] Bochev, P.B., Dohrmann, C.R., Gunzburger, M.D.: Stabilization of low-order mixed finite elements for the Stokes equations. SIAM J. Numer. Anal. 44, 82–101 (2006)

[2] Brezzi, F.: A discourse on the stability conditions for mixed finite element formulations. Comput. Methods Appl. Mech. Eng. 82, 27–57 (1990)

[3] Liu, F., Borja, R.I.: Stabilized low-order finite elements for frictional contact with the extended finite element method. Comput. Methods Appl. Mech. Eng. 82, 27–57 (2010)

[4] White, J.A., Borja, R.I.: Stabilized low-order finite elements for coupled solid-deformation/fluid-diffusion and their application to fault zone transients. Comput. Methods Appl. Mech. Eng. 199, 2456–2471 (2008)

[5] White, J.A.: Stabilized finite element methods for coupled flow and geomechanics. Ph.D. Thesis, Stanford University (2009)

EXPERIMENTAL AND COMPUTATIONAL ISSUES IN THE MECHANICS OF MULTI-PHYSICAL UNSATURATED SOIL

W. Ehlers and O. Avci

Institute of Applied Mechanics (CE)
University of Stuttgart
Pfaffenwaldring 7, 70569 Stuttgart, Germany
e-mail: {ehlers,avci}@mechbau.uni-stuttgart.de,
web page: http://www.mechbau.uni-stuttgart.de/ls2

Summary. The article concerns the combination of unsaturated soil mechanics with shear band initiation phenomena. Based on the Theory of Porous Media, the soil model proceeds from an elasto-plastic set-up based on monotonic and non-monotonic homogeneous as well as inhomogeneous shear-band localisation experiments.

Keywords: finite element computations, laboratory tests, porous media mechanics, unsaturated soil.

1 INTRODUCTION

Computational investigations of complex multi-physical problems in soil mechanics become more and more important due to an increasing number of coupled solid deformation and pore-fluid flow problems as well as various instability phenomena induced by variations of the groundwater table initiated by rainfall events. Laboratory experiments and numerical simulations can support the detection of coupled failure mechanisms by changing the initial and boundary conditions of the considered experiment and the corresponding numerical computation formulated as an initial-boundary-value problem (IBVP) within the finite-element method (FEM). This procedure can contribute to a deeper understanding of various problems such as complex slope failure processes. In the present article, the soil system, generally understood as an unsaturated soil, is described as a triphasic multi-physical material consisting of a soil skeleton, such as fine sand, a pore liquid, such as water, and a pore gas, such as air. The model is embedded in the well-founded Theory of Porous Media (TPM), while numerical solutions are realised by use of the finite-element solver PANDAS. Proceeding from cohesionless fine sand as the basic soil under study, the material parameters governing the soil behaviour are taken from triaxial experiments on dry sand specimens carried out under homogeneous loading conditions, whereas the parameters governing the hydraulic behaviour are determined by experiments on saturated soil specimens under non-deforming conditions.

The experimental evidence on well-defined GEBA fine sand[1] of the Degebo[2] revealed that the standard elasto-plasticity model of soil materials consisting of a

[1] Sand from the Gebenbacher sand pit (trade name GEBA, Dorfner, Hirschau, Germany) with the following characteristics: grain size: 0.03 – 0.3 mm; sieve retention: d_{10}=0.09 mm, d_{60}=0.11 mm, degree of uniformity: Cu = 1.22.

[2] **De**utsche Forschungs**ge**sellschaft für **Bo**denmechanik (German Scientific Society for Soil Mechanics).

R.I. Borja (Ed.): Multiscale and Multiphysics Processes in Geomechanics, SSGG, pp. 125–128.
springerlink.com

geometrically linear approach based on the Hookean elasticity concept and a hardening plasticity scheme in the framework of plastic work contours in connection with a constant failure surface is insufficient to describe inhomogeneous and non-monotonic experiments including failure mechanisms. As a result, this article does not only include the description of unsaturated soil, in general, but also aims at presenting a new elasto-plasticity approach for sandy soil.

Finally, a numerical investigation of a slope failure scenario of an unsaturated soil is presented revealing the strong coupling of the soil deformation and the hydraulic behaviour during failure processes.

2 THE UNSATURATED SOIL MODEL

In the present contribution, the triphasic unsaturated soil model is assumed to consist of a materially incompressible, elasto-plastic or elasto-viscoplastic, frictional solid skeleton with the pore space filled by a combination of two viscous fluids, namely, a materially incompressible pore liquid and a materially compressible pore gas. Based on the property of immiscible materials together with the principle of phase separation [1], the behaviour of the components can be studied separately and put together further on by adding the coupling mechanisms. Consequently, the experiments for the parameter identification process can be carried out independently for the solid deformation behaviour and for the hydraulic properties. In particular, the solid matrix response can be observed and determined from triaxial tests carried out under dry and homogeneous conditions, whereas the hydraulic properties, such as the capillary-pressure-saturation relation and the relative permeabilities governing the partially saturated zone, can be measured without exerting an external load onto the sand specimen.

2.1 Basic Equations

The governing equations of the model under consideration are briefly presented within the framework of the TPM, cf. [2, 3, 4] and the quotations therein. The weak formulation of the overall momentum balance yields

$$\int_{\Omega}(\boldsymbol{\sigma}_E^S - p\mathbf{I})\cdot \operatorname{grad}\delta\mathbf{u}_S \,\mathrm{d}v = \int_{\Omega}\rho\mathbf{g}\cdot\delta\mathbf{u}_S\,\mathrm{d}v + \int_{\Gamma_t}\bar{\mathbf{t}}\cdot\delta\mathbf{u}_S\,\mathrm{d}a\,, \tag{1}$$

where $\boldsymbol{\sigma}_E^S$ is the solid extra (effective) stress, $\mathbf{I}$ is the identity tensor and p is the total excess pore pressure given by the sum of the saturation-weighted effective fluid pressures of the pore water, p^{WR}, and pore air, p^{AR}:

$$p = s^W p^{WR} + s^A p^{AR}\,. \tag{2}$$

Moreover, ρ is the so-called overall density, $\mathbf{g}$ is the mass-specific gravitation, $\bar{\mathbf{t}} = (\boldsymbol{\sigma}_E^S - p\mathbf{I})\mathbf{n}$ is the external load vector acting on the *Neumann* boundary Γ_t of the overall medium and '$\operatorname{grad}(\cdot)$' denotes the gradient operator.

The weak formulation of the volume balance of the materially incompressible pore water with constant effective density ($\rho^{WR} = \text{const.}$) is given by

$$\int_{\Omega}[(n^W)'_S + n^W \operatorname{div}(\mathbf{u}_S)'_S]\delta p^{WR}\,\mathrm{d}v - \int_{\Omega} n^W \mathbf{w}_W \cdot \operatorname{grad}\delta p^{WR}\,\mathrm{d}v = -\int_{\Gamma_v}\bar{v}^W \delta p^{WR}\,\mathrm{d}a\,, \tag{3}$$

while the mass balance of the compressible pore air reads

$$\int\limits_{\Omega}[(n^A\rho^{AR})'_S + n^A\rho^{AR}\mathrm{div}(\mathbf{u}_S)'_S]\,\delta p^{AR}\,\mathrm{d}v - \int\limits_{\Omega} n^A\rho^{AR}\mathbf{w}_A\cdot\mathrm{grad}\,\delta p^{AR}\,\mathrm{d}v = -\int\limits_{\Gamma_q}\bar{q}^A\,\delta p^{AR}\,\mathrm{d}a\,. \tag{4}$$

Therein, $\bar{v}^W = n^W\mathbf{w}_W\cdot\mathbf{n}$ is the efflux of liquid volume through the *Neumann* boundary Γ_v with the outward-oriented unit surface normal $\mathbf{n}$, while $\bar{q}^A = n^A\rho^{AR}\mathbf{w}_A\cdot\mathbf{n}$ is the efflux of gaseous mass through the *Neumann* boundary Γ_q. Moreover, '$\mathrm{div}(\cdot)$' is the divergence operator corresponding to '$\mathrm{grad}(\cdot)$' and $(\cdot)'_S$ indicates the material time derivative with respect to the solid motion. In addition, the relative fluid velocities are determined from *Darcy*-type filter laws

$$n^\beta\mathbf{w}_\beta = -\frac{K_r^\beta}{\gamma^{\beta R}}(\mathrm{grad}\,p^{\beta R} - \rho^{\beta R}\mathbf{g})\,, \tag{5}$$

where $\gamma^{\beta R}$ is the specific weight of the fluid φ^β, pore water or pore air, and K_r^β is the corresponding relative *Darcy* permeability.

2.2 The Sandy Soil Model

The assumption of rigid and incrushable soil grains requires that the elastic *Helmholtz* free-energy function of the solid skeleton must increase to infinity under hydrostatic compression. In order to take this into account, Ehlers and Scholz [5] proposed a materially non-linear elastic soil model of the form:

$$\boldsymbol{\sigma}_E^S = 2\mu^S\boldsymbol{\varepsilon}_{Se} + \lambda^S\left(\frac{\varepsilon^V_{Se,\,crit}}{\varepsilon^V_{Se,\,crit} - \varepsilon^V_{Se}}\right)\varepsilon^V_{Se}\,\mathbf{I}\,. \tag{6}$$

Therein, μ^S and λ^S are the macroscopic *Lamé* constants, $\boldsymbol{\varepsilon}_{Se}$ is the elastic part of the linear strain tensor, $\varepsilon^V_{Se} = \boldsymbol{\varepsilon}_{Se}\cdot\mathbf{I}$ is the elastic volume strain and $\varepsilon^V_{Se,\,crit}$ its critical value that cannot be exceeded under compressive loads.

The plastic part of the material behaviour is based on the single-surface yield function [6].

$$F(\boldsymbol{\sigma}_E^S) = \sqrt{\mathrm{II}^{\mathrm{D}}_\sigma(1+\gamma\,\mathrm{III}^{\mathrm{D}}_\sigma/(\mathrm{II}^{\mathrm{D}}_\sigma)^{3/2})^m + \tfrac{1}{2}\alpha\,\mathrm{I}^2_\sigma + \delta^2\mathrm{I}^4_\sigma} + \beta\,\mathrm{I}_\sigma + \varepsilon\,\mathrm{I}^2_\sigma - \kappa = 0. \tag{7}$$

In this equation, I_σ is the first principal invariant of $\boldsymbol{\sigma}_E^S$, while $\mathrm{II}^{\mathrm{D}}_\sigma$ and $\mathrm{III}^{\mathrm{D}}_\sigma$ are the (negative) second and third principal deviatoric invariants of $\boldsymbol{\sigma}_E^S$. Furthermore, $\boldsymbol{S}_h = \{\alpha,\ \beta,\ \delta,\ \varepsilon,\ \kappa\}$ and $\boldsymbol{S}_d = \{\gamma,\ \boldsymbol{m}\}$ are sets of hydrostatic and deviatoric material parameters that have to be determined from experiments revealing hardening as well as softening soil properties and varying failure surfaces, cf. Ehlers et al. [7].

3 NUMERICAL EXAMPLE

The numerical example presented in this article exhibits a slope with water tables on both sides. Before loading by q, both water tables have been set to 10 cm. Afterwards, q has been increased until a shear occurs. Figure 1 presents the basic geometry and the water saturation of the undeformed soil as a result of constant water tables on both sides. Note that, as a result of the fine-sand capillarity, the saturated zone reaches up to approximately 60 cm from the bottom before the partially saturated zone starts. At slope failure, cf. Figure 2, the fully saturated zone decreases rapidly as a result of dilatancy in the shearing zone. It is furthermore observed that the streamlines of the pore water are also pointing towards the shear bands.

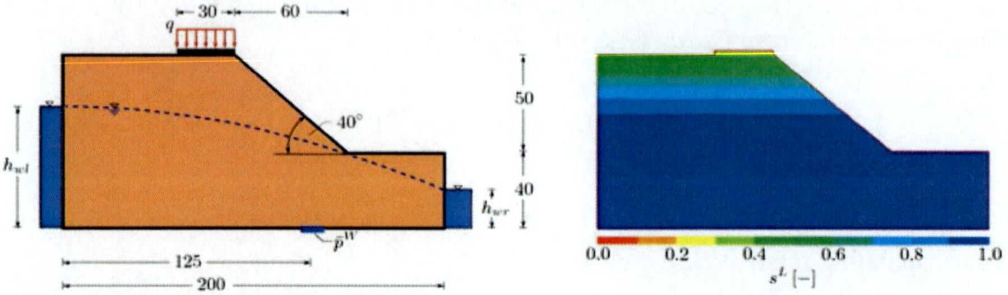

Fig. 1. Geometry and dimensions of the slope in cm (left) and computed water saturation for a water table of $h_{wl} = h_{wr} = 10\,\text{cm}$ (right).

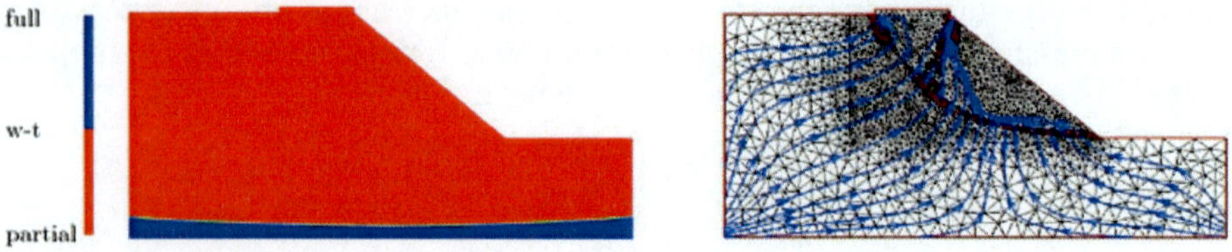

Fig. 2. Water table at slope failure (left), and streamlines of water seepage flow and shear band (right).

REFERENCES

[1] Ehlers, W.: On thermodynamics of elasto-plastic porous media. Archive of Mechanics 19, 73–93 (1989)

[2] Ehlers, W.: Foundations of multiphasic and porous materials. In: Ehlers, W., Bluhm, J. (eds.) Porous Media: Theory, Experiments and Numerical Applications, pp. 3–86. Springer, Berlin (2002)

[3] Ehlers, W., Graf, T., Ammann, M.: Deformation and localization analysis in partially saturated soil. Computer Methods in Applied Mechanics and Engineering 193, 2885–2910 (2004)

[4] Borja, R.I., Lai, T.Y.: Propagation of localization instability under active and passive loading. Journal of Geotechnical and Geoenvironmental Engineering, ASCE 128, 64–75 (2002)

[5] Ehlers, W., Scholz, B.: An inverse algorithm for the identification and the sensitivity analysis of the parameters governing micropolar elasto-plastic granular material. Archive of Applied Mechanics 77, 911–931 (2007)

[6] Ehlers, W.: A single-surface yield function for geomaterials. Archive of Applied Mechanics 65, 246–259 (1995)

[7] Ehlers, W., Avci, O., Markert, B.: Computation of slope movements initiated by rain-induced shear bands in small-scale tests and in situ. Vadoze Zone Journal (2010) (accepted)

SCALABLE PRECONDITIONING TECHNIQUES FOR FULLY-COUPLED HYDROMECHANICAL MODELS

Joshua A. White[1] and Ronaldo I. Borja[2]

[1] Computational Geosciences Group
Lawrence Livermore National Laboratory
P.O. Box 808, L-231, Livermore, CA, 94551, USA
e-mail: jawhite@llnl.gov
[2] Civil and Environmental Engineering
Stanford University
Stanford, CA, 94305, USA
e-mail: borja@stanford.edu

Summary. In this study, we consider efficient solution methods for mixed finite element models of variably-saturated fluid flow through deformable porous media. Our main focus is preconditioning techniques to accelerate the convergence of implicit Newton–Krylov solvers. We highlight an approach in which preconditioners are built from block-factorizations of the coupled system. The resulting methodology allows one to extend single-physics preconditioners in a natural way to multi-physics applications, allowing for significant code reuse while still capturing the tightly-coupled nature of the underlying physics. The proposed methodology has been extensively tested on large, three-dimensional problems with several million unknowns. Results demonstrate that an algebraic multigrid variant of the block-preconditioner leads to mesh-independent solver convergence and good parallel efficiency.

Keywords: algebraic multigrid, coupled geomechanics, finite elements, Newton-Krylov.

1 INTRODUCTION

The focus of this study is efficient numerical methods for modeling variably-saturated geomaterials. In particular, we examine fully-coupled models that simultaneously capture the deformation of the solid matrix and fluid flow through the pore space. In many geotechnical and geoscientific applications, it is necessary to model this fluid–structure interaction in a tightly-coupled way to make meaningful predictions. Unfortunately, these multiphysics models are challenging—from both a theoretical and numerical point of view—and their applications in day-to-day practice have been limited. We attempt to address some of these challenges here, particularly as to how one can design efficient and scalable algorithms for this class of problems.

To fix ideas, consider a basic model for the behavior of variably saturated geomaterials. The mixture response is governed by two coupled equations: a mass balance and a linear momentum balance:

$$\theta\dot{\psi} + \psi\nabla\cdot\dot{\boldsymbol{u}} + \nabla\cdot\boldsymbol{w} = 0. \tag{1}$$

$$\nabla\cdot(\boldsymbol{\sigma}' - \psi p\mathbf{1}) + \rho\boldsymbol{g} = \mathbf{0}. \tag{2}$$

R.I. Borja (Ed.): Multiscale and Multiphysics Processes in Geomechanics, SSGG, pp. 129–132.
springerlink.com

where, θ is the porosity, ψ is the water-phase saturation, $\dot{\boldsymbol{u}}$ is the solid displacement velocity, $\boldsymbol{w}$ is the seepage (Darcy) velocity, $\boldsymbol{\sigma}'$ is the effective stress, p is the water phase pressure, $\mathbf{1}$ is a second-order unit tensor, and $\rho\boldsymbol{g}$ is a body-force due to self-weight (1). The governing equations are supplemented with the usual boundary and initial conditions, as well as several constitutive relationships for the solid and fluid behaviors.

There are a variety of ways to discretize the above model, but here we examine a two-field $\boldsymbol{u}/p$ formulation, in which the solid displacement and water pressure are chosen as primary unknowns. Following the standard methodology, the finite element discretization leads to a nonlinear system of residual equations that must be solved to advance the solution in each time-step of the simulation. The roots of these residual equations are found using Newton's method, and it is this portion of the code that dominates the computational expense of a typical simulation. Because the coupled formulation is a multi-field problem, the Jacobian system that must be solved in each Newton iteration inherits a useful block-structure:

$$\begin{bmatrix} A & B_1 \\ B_2 & C \end{bmatrix} \begin{bmatrix} \Delta U \\ \Delta P \end{bmatrix} = - \begin{bmatrix} R_{\text{mom.}} \\ R_{\text{mass}} \end{bmatrix} \tag{3}$$

Unfortunately, this system is typically ill-conditioned, making its solution using iterative solvers inefficient without good preconditioning. That is, instead of solving the original linear system $Jx = b$ using an iterative solver, we can construct a preconditioning operator P and instead solve the (left) preconditioned system $P^{-1}Jx = P^{-1}b$. If the matrix $P^{-1}J$ has better spectral properties than J itself, then the Krylov solver will converge in fewer iterations.

Roughly speaking, our aim here is to design preconditioners that are informed of the coupled nature of the underlying problem and can exploit it to achieve good convergence properties. To this end, we make the preliminary observation that the Jacobian admits a variety of block-factorizations (2). For example, a block LU factorization is

$$J = \begin{bmatrix} A & 0 \\ B_2 & S \end{bmatrix} \begin{bmatrix} I & A^{-1}B_1 \\ 0 & I \end{bmatrix} \tag{4}$$

where $S = C - B_2 A^{-1} B_1$ is the Schur-complement (of A) for the system. This decomposition suggests an effective left-acting preconditioner would be (3),

$$P = \begin{bmatrix} A & 0 \\ B_2 & S \end{bmatrix} \tag{5}$$

which implies,

$$P^{-1} = \begin{bmatrix} A^{-1} & 0 \\ -S^{-1}B_2A^{-1} & S^{-1} \end{bmatrix} \quad \text{and} \quad P^{-1}J = \begin{bmatrix} I & A^{-1}B_1 \\ 0 & I \end{bmatrix} \tag{6}$$

We observe that the preconditioned Jacobian $P^{-1}J$ is block upper triangular, with identity blocks on the diagonal. It is straightforward to show that the preconditioned matrix has only one distinct eigenvalue ($\lambda = 1$), and a Krylov-based iteration on the preconditioned system would converge in at most two iterations. In practice, however, it is too expensive to form the "exact" preconditioner. Note, for example, that the Schur-complement S is dense (due to the presence of A^{-1}) and is therefore difficult to work with. Also, we do not want to explicitly form any of the inverse operators appearing in (6). Instead, we replace all of these operators with their own "sub-preconditioners," which by definition are good approximate inverses,

$$\tilde{P}^{-1} = \begin{bmatrix} P_A^{-1} & 0 \\ -P_S^{-1} B_2 P_A^{-1} & P_S^{-1} \end{bmatrix} \tag{7}$$

A block-based strategy now emerges. First, we compute sub-preconditioners P_A and P_S for A and S respectively, and then combine them into a global preconditioner $\tilde{P}$. Of course, the success of this strategy depends on developing effective sub-preconditioners, and is therefore a major focus of this study. In comparing several strategies, the best performance was achieved using an Algebraic Multigrid method for the sub-preconditioners. We refer the interested reader to (4) for further details, particularly on how best to precondition the Schur-complement operator for this class of problems.

2 APPLICATIONS

The performance of the preconditioning strategy has been tested on a variety of linear and nonlinear problems. For example, Table 1 presents the results of a weak scaling study performed on a geotechnical footing test problem. The soil is modeled as saturated and linear elastic. In this limiting case, the problem is linear in each time step, and Newton's method converges in a single iteration. The problem therefore provides a way to test the performance of the preconditioner independently of the outer Newton-solver. The problem is formulated using equal-order Q1/Q1 hexahedra, with a stabilizing modification to deal with inf–sup instabilities (5). We observe that the number of iterations per solve remains essentially constant regardless of the domain size. Furthermore, the total wall time remains almost constant, except for a small overhead associated with the parallel nature of the computation. We see that the preconditioned Krylov solver shows good scaling and parallel efficiency, and can solve fully-coupled systems with several million degrees of freedom in a few seconds. This fact is crucial as we consider nonlinear problems, because a typical time-dependent, nonlinear solution may require computing hundreds or thousands of linear Newton updates.

Figure 1 shows one such application, modeling the geomechanical response of a CO_2 reservoir during injection. Here, it is necessary to discretize the full three-dimensional geometry of the reservoir, overburden, and underburden. The situation is also complicated by several cross-cutting faults that can either enhance the flow or serve as compartmentalizing barriers. As the faults are pressured, they may also dilate and slip. Modeling these interactions clearly requires both a tightly coupled model and highly-efficient, scalable solution methods.

Table 1. Parallel scaling of an algebraic multigrid variant of the block preconditioner.

Processors	1	4	16	64
Elements	24,389	97,556	390,224	1,560,896
Degrees of Freedom	55,296	417,720	1,642,680	6,514,680
Avg. Iterations	12.1	13.1	12.5	11.7
Avg. Time (s)	9.87	10.84	10.85	11.28

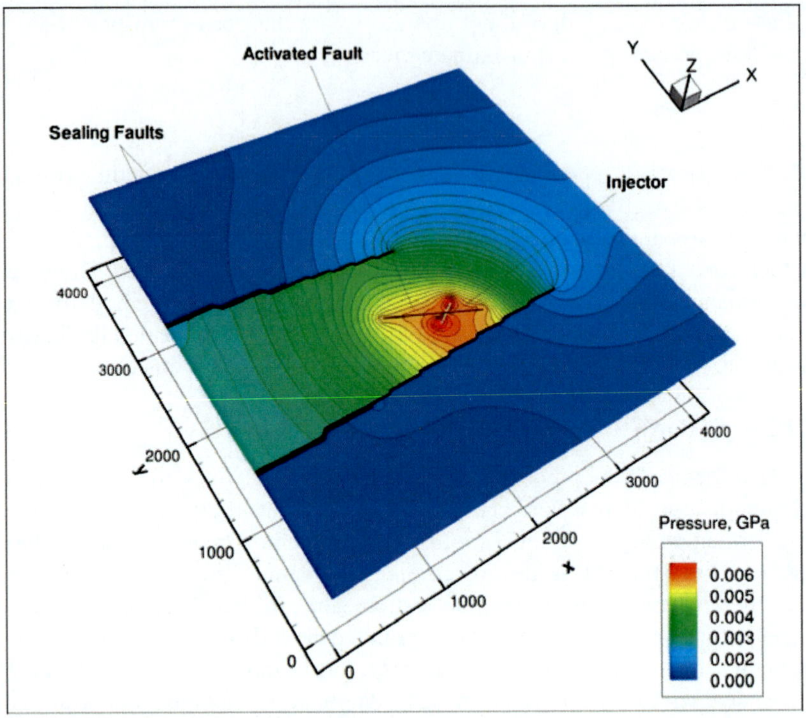

Fig. 1. Pressurization of faults due to CO_2 injection into a reservoir.

ACKNOWLEDGEMENTS

This study was performed under the auspices of the U.S. Department of Energy by Lawrence Livermore National Laboratory under Contract DE-AC52-07NA27344. The first author is grateful for the support extended by the Lawrence Postdoctoral Fellowship Program.

REFERENCES

[1] Borja, R.I.: On the mechanical energy and effective stress in saturated and unsaturated porous continua. International Journal of Solids and Structures 43(6), 1764–1786 (2006)
[2] Benzi, M., Golub, G.H., Liesen, J.: Numerical solution of saddle point problems. Acta Numerica 14, 1–137 (2005)
[3] Bramble, J.H., Pasciak, J.E.: A preconditioning technique for indefinite systems resulting from mixed approximation of elliptic problems. Math. Comput. 50(181), 1–17 (1988)
[4] White, J.A., Borja, R.I.: Block-preconditioned Newton-Krylov solvers for fully-coupled flow and geomechanics. Computational Geosciences (2011) (in press)
[5] White, J.A., Borja, R.I.: Stabilized low-order finite elements for coupled solid-deformation/fluid-diffusion and their application to fault zone transients. Computer Methods in Applied Mechanics and Engineering 197(49-50), 4353–4366 (2008)

MODELING MULTI-SCALE FLOW USING THE MATERIAL POINT METHOD

Pedro Arduino*, Peter Mackenzie-Helnwein, and Gregory R. Miller

Department of Civil and Environmental Engineering
201 More Hall, Box 352700
University of Washington, Seattle, WA 98195-2700
*e-mail: parduino@uw.edu, web page: http://www.ce.washington.edu

Summary. We demonstrate that the multi-linear interpolation functions used on an Eulerian grid in standard MPM generate volumetric locking, which dominates the solution as bulk compressibility approaches the incompressible limit. We propose a simple relaxation procedure based on a three-field Hu-Washizu principle and demonstrate its applicability to modeling of granular flow problems.

Keywords: granular flow, meshfree methods, MPM, volumetric locking.

1 INTRODUCTION

While granular flows can extend many meters in depth and more than 100 meters in length, the controlling mechanisms have length scales on the order of grains (in sands) or boulders (in debris-flows). Modeling challenges emerge at all scales, governed by limitations on computational resources and/or limitations of the numeric techniques employed at any particular length scale.

In this article, we explore the suitability and reliability of the Material Point Method (MPM) for modeling granular flow problems at the largest, phenomenological scales of consideration. The MPM is a numeric technique that has its roots in fluid mechanics and was extended for use in solid mechanics by Sulsky et al. [1]. In its common form, MPM utilizes continuum-based constitutive relations to represent material of a finite volume, called material points or particles, using a single set of state variables to represent the homogenized state of the associated material. In this article, behavior at smaller scales is represented by reasonable rheological models at the macro-scale. This is not a requirement for the present discussion, but it rather removes concerns associated with numeric homogenization procedures or deficiencies in a smaller scale model.

2 APPROACH

Granular material at small deformations can show significant dilative behavior [2]. For the large deformations observed in debris flows, dilation quickly reaches a limit state, and subsequent flow needs to be treated as incompressible in an average sense. This incompressibility imposes a challenge on any numeric technique employed at the macroscopic level, including the MPM.

A benefit of the MPM frequently referenced in the literature is its ability to dynamically model large deformations of heterogeneous three-dimensional bodies. This clear benefit over the Finite Element Method (FEM) is a consequence of a mesh-free representation of the body through particles. The equation of motion is solved in a weak

R.I. Borja (Ed.): Multiscale and Multiphysics Processes in Geomechanics, SSGG, pp. 133–136.
springerlink.com

form using continuous interpolation functions on a regular Eulerian background grid. These interpolation functions, however, provide a close tie to the FEM and, thus, the MPM inherits several benefits but also some problems derived from the FEM.

2.1 Volumetric Locking

While the MPM does not suffer from mesh distortion, Shin [3] clearly demonstrates that the MPM can exhibit volumetric locking. This can be illustrated by applying the MPM to the problem of a dam-break problem with fluid as the material. In this case, the material is characterized using a free energy function:

$$\psi(\boldsymbol{\varepsilon}, \boldsymbol{\xi}) = \bar{\psi}(\operatorname{dev} \boldsymbol{\varepsilon}, \boldsymbol{\xi}) + U(\theta) \tag{1}$$

where $\boldsymbol{\varepsilon}$ is the Almansi strain tensor, $\boldsymbol{\xi}$ is the suitable set of history dependent variables, $\operatorname{dev} \boldsymbol{\varepsilon}$ is the deviatoric strain, and $\theta = \operatorname{tr} \boldsymbol{\varepsilon}$ is the volumetric strain. $U(\theta)$ is the volumetric portion of the free energy. The stress–strain relation can be expressed as

$$\boldsymbol{\sigma} = \varrho \frac{\partial \psi}{\partial \boldsymbol{\varepsilon}} + 2\mu\, \boldsymbol{d} = \varrho \operatorname{dev} \left(\frac{\partial \bar{\psi}}{\partial \operatorname{dev} \boldsymbol{\varepsilon}} \right) + \varrho \frac{\partial U(\theta)}{\partial \theta} \mathbf{1} + 2\mu\, \boldsymbol{d} \tag{2}$$

with μ as dynamic viscosity and $\boldsymbol{d} = (\nabla \boldsymbol{v} + \nabla \boldsymbol{v}^T)/2$ as the rate of deformation tensor. $\boldsymbol{u}(\boldsymbol{x})$ is the displacement field, and $\boldsymbol{v} = \dot{\boldsymbol{u}}$ is the velocity. The strain gradients in (2) represent history-dependent portions of stress. The remainder is a viscous term representing the rate-dependent but history-independent portion of stress.

Letting $\bar{\psi} = G \operatorname{dev} \boldsymbol{\varepsilon} : \operatorname{dev} \boldsymbol{\varepsilon}$, $U = k\theta^2/2$, and $\mu = 0$ results in a linear elastic material, while choosing $\bar{\psi} = 0$ and $\mu > 0$ results in a (nearly incompressible) Newtonian fluid. Transition states are obtained for $\bar{\psi} \neq 0$, $U(\theta) \neq 0$, and $\mu > 0$.

Figure 1 shows a plane dam-break problem in a finite domain that is modeled using $\bar{\psi} = 0$ and $U(\theta) = k\theta^2/2$ for a large bulk modulus $k = 2.0$ GPa and $\mu = 10^{-6}$ kPa, both typical for water at room temperature. Figure 1a shows the initial hydrostatic state of stress, which is properly represented. However, using standard MPM with the nearly incompressible constitutive model results in a quasi-static state shown in Figure 1b. This state is reached after just 1.4 seconds. The fluid never reaches the right container wall. Pressure varies excessively between particles, and peak values are about two orders of magnitude above the expected.

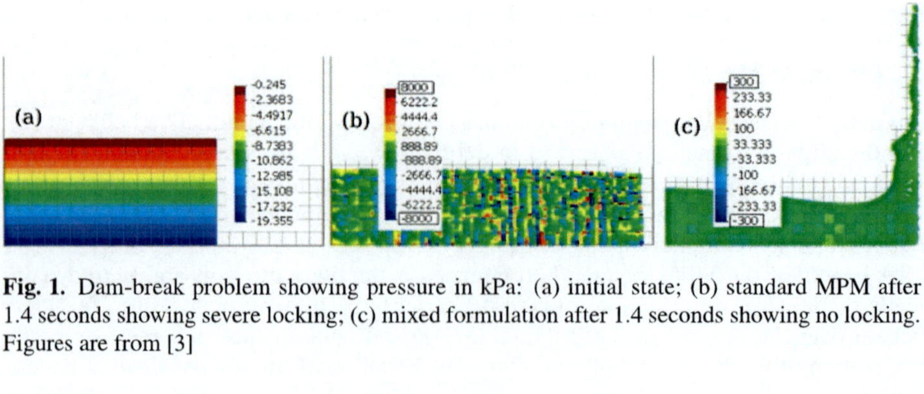

Fig. 1. Dam-break problem showing pressure in kPa: (a) initial state; (b) standard MPM after 1.4 seconds showing severe locking; (c) mixed formulation after 1.4 seconds showing no locking. Figures are from [3]

The source of the problem primarily lies in the way the volumetric strain rate $\dot{\theta}$ is computed from the standard MPM multi-linear interpolation functions, $\phi_i(x, y, z)$. Representing a nearly incompressible flow using quadrilaterals (2D) or 8-node bricks (3D), in general, does not properly reproduce a constant volumetric strain rate throughout the cell. Instead, an artificial variation of $\dot{\theta}$ generates artificial variations in pressure. Particles crossing into adjacent cells add to the generation of numeric oscillations [4], as is clearly visible in Figure 1b.

2.2 Mitigating Volumetric Locking

A suitable solution for the FEM equivalent problem was proposed by Weissman and Taylor [5] and Liu et al. [6]. They use a Hu-Washizu variational principle to introduce a piecewise smooth dilation, $\bar{\theta}$, and pressure, $\bar{p}$, and tie them in a weak form to the respective variables derived from the interpolated displacement field. Using the regular background grid in MPM, the respective quasi-static equations for the Hu-Washizu type solution simplify to

$$\int_V \varrho\bar{\psi}(\operatorname{dev}\boldsymbol{\varepsilon},\boldsymbol{\xi})\,dV + \int_V \varrho U(\bar{\theta})\,dV - \int_V (\bar{\theta} - \operatorname{tr}\boldsymbol{\varepsilon})\bar{p}\,dV - \int_V \varrho\boldsymbol{g}\cdot\boldsymbol{u}\,dV \rightarrow \text{stationary}\,, \tag{3}$$

where, for simplicity, only gravitational loads are included. Variation with respect to the independent variables $\boldsymbol{u}$, $\bar{\theta}$, and $\bar{p}$ and integration by parts of all terms including derivatives of the variational displacement, $\delta\boldsymbol{u}$, yields the Euler equations. In addition to the equation of motion, we obtain

$$\bar{\theta} = \operatorname{div}\boldsymbol{u}\,, \qquad \bar{p} = \varrho\frac{\partial U(\bar{\theta})}{\partial\bar{\theta}}\,, \qquad \text{and} \quad \boldsymbol{\sigma} = \varrho\operatorname{dev}\left(\frac{\partial\psi}{\partial\operatorname{dev}\boldsymbol{\varepsilon}}\right) + \bar{p}\,\mathbf{1}\,. \tag{4}$$

An efficient technique is obtained by choosing $\bar{p}$, and hence $\delta\bar{p}$, in (3) to be constant per cell, c, but zero outside that cell. Converting volume integrals to mass integrals and computing them numerically as sums over particle domains yield the MPM-compatible cell-averaged volumetric strain and pressure, respectively, as

$$\bar{\theta}_c = \sum_{p\in c}\operatorname{tr}\boldsymbol{\varepsilon}_p\frac{m_p}{\varrho_p}\Big/\sum_{p\in c}\frac{m_p}{\varrho_p} \quad \text{and} \quad \bar{p}_c = \varrho_c\frac{\partial U(\bar{\theta}_c)}{\partial\bar{\theta}_c}\,, \tag{5}$$

in which ϱ_c is a cell-averaged density. Equation (5) is evaluated as an additional initial step before computing the MPM algorithm for a new time step. In order to preserve material history, particle strain and stress, respectively, are modified as

$$\boldsymbol{\varepsilon}_p \rightarrow \operatorname{dev}\boldsymbol{\varepsilon}_p + \frac{\bar{\theta}_c}{3}\mathbf{1} \quad \text{and} \quad \boldsymbol{\sigma}_p \rightarrow \operatorname{dev}\boldsymbol{\sigma}_p + \bar{p}_c\,\mathbf{1} \tag{6}$$

for all particles in cell c.

Adding this simple procedure to the standard algorithm restores virtually locking-free and physically correct fluid behavior, as shown in Figure 1c.

Granular flow differs from the above fluid flow problem by the presence of shear stiffness through a nonvanishing $\bar{\psi}(\boldsymbol{\varepsilon},\boldsymbol{\xi})$ and limited shear strength as a function of the acting mean pressure. The present study utilizes a Drucker-Prager type model for the granular material – solely for isolating the locking and the cell-crossing problem. History variables $\boldsymbol{\xi}$ consist of the plastic strain tensor only.

Figure 2 shows the simulation of a sand column collapse. The problem involves a static initial stress state, dynamic collapse, and a final state, again at rest. Rapid removal

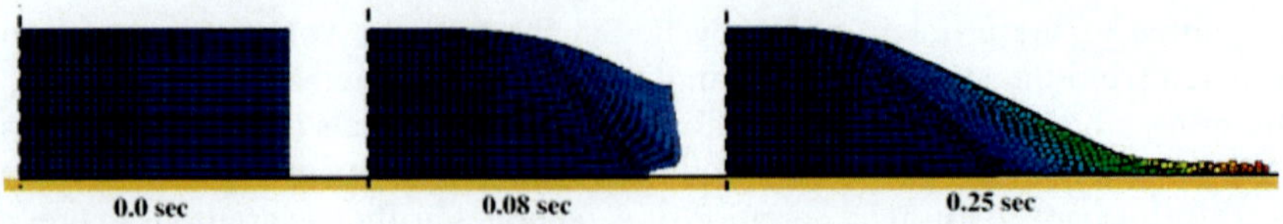

Fig. 2. Plane collapse of a sand column. From the left: Initial configuration; Intermediate state; Final configuration.

of a supporting wall triggers slope failure close to the theoretical active failure mode. The failed material is deposited at the foot of the sand column, and a final state slightly shallower than the static angle of repose develops. For further details on this simulation, including a highly satisfactory comparison to experimental data, see Shin [3].

A similar simulation using standard MPM produced the initial slope failure, but could not produce the final deposit geometry. Note that even a finite deformation analysis using the FEM cannot simulate this process without repeated remeshing. The MPM with the proposed mixed formulation, however, does not require any modification of the underlying regular Eulerian grid, nor any special treatment at the particle level to generate this result.

3 CONCLUSIONS

The modified MPM approach presented here proves to be simple and effective. In addition to removing volumetric locking due to the low order of the underlying interpolation functions (FEM similarity), the presented technique also significantly reduces the MPM-specific cell-crossing error. The computational cost for the presented anti-locking procedure is negligible in comparison to the standard MPM algorithm, even if finite strain constitutive models are employed.

REFERENCES

[1] Sulsky, D., Zhou, S., Schreyer, H.L.: Application of a particle-in-cell method to solid mechanics. Computer Physics Communications 87(1-2), 236–252 (1995)

[2] Li, X.S., Dafalias, Y.F.: Dilatancy for cohesionless soils. Géotechnique 50(4), 449–460 (2000)

[3] Shin, W.K.: Modeling Mixing and Separation of Solid Matter and Fluid in Landslides and Debris Flows by Representing the Multiphase Material through Distinct Phases. PhD thesis, University of Washington, Seattle, WA (2009)

[4] Bardenhagen, S.G., Kober, E.M.: The generalized interpolation material point method. Computer Modeling in Engineering and Sciences 5(6), 477–495 (2004)

[5] Weissman, S.L., Taylor, R.L.: Treatment of internal constraints by mixed finite element methods: Unification of concepts. Int. J. Num. Meth. Engng. 33, 131–141 (1992)

[6] Liu, C.H., Hofstetter, G., Mang, H.A.: 3D finite element analysis of rubber-like materials at finite strains. Engineering Computations 11(2), 111–128 (1994)

A NUMERICAL MODEL FOR GEOTECHNICAL PROBLEMS INVOLVING PARTIALLY SATURATED SOILS

G. Hofstetter*, M. Pertl, and M. Hofmann

Department of Civil Engineering
University of Innsbruck
Technikerstraße 13, A-6020 Innsbruck, Austria
*e-mail: Guenter.Hofstetter@uibk.ac.at, web page: http://www.uibk.ac.at/bft/

Summary. The paper deals with a numeric model for coupled solid–fluid FE analyses of geotechnical problems involving partially saturated soils. The Barcelona Basic Model serves as a constitutive model for the soil skeleton. Different stress update algorithms are compared for this model regarding robustness, accuracy, and efficiency. The application of the numerical model is demonstrated by numerical simulations of the impoundment of an earth dam and of the injection of compressed air into water saturated soil.

Keywords: multi-phase model, partially saturated soil, unsaturated soil model.

1 INTRODUCTION

A finite element model for partially saturated soils relies on a FE formulation for three-phase media, consisting of a deformable soil skeleton and the fluid phases: water and air [1], and on a constitutive model for the deformable soil skeleton, which allows describing essential features of the behavior of partially saturated soils. In the present FE model the displacements $\mathbf{u}$ of the soil skeleton, the capillary pressure p^c, and the air pressure p^a are chosen as primary variables. A Newton-type solution procedure is employed for solving the system of nonlinear coupled equations of the three-phase model.

As constitutive model, the well-known Barcelona Basic Model [2] formulated in terms of net stress and capillary pressure, is chosen. Alternatively, for partially saturated sands and silts a cap model, which is based on the generalized effective stress and capillary pressure, can be employed [3]. Within the framework of a FE model, the stress update algorithm for the partially saturated soil model plays an important role. In this paper, different stress update algorithms, consisting of (i) an explicit stress update algorithm, (ii) a general return mapping algorithm, (iii) an optimized return mapping algorithm, and (iv) a semi-explicit stress update algorithm are compared with respect to robustness, accuracy, and efficiency. The performance of the stress update algorithms is shown in a systematic manner for a large range of prescribed combinations of volumetric and deviatoric strain increments.

Finally, the application of the developed coupled FE model is demonstrated through coupled transient numerical simulations of geotechnical problems involving partially saturated soils.

R.I. Borja (Ed.): Multiscale and Multiphysics Processes in Geomechanics, SSGG, pp. 137–140.
springerlink.com

2 STRESS UPDATE ALGORITHMS

Within the framework of a FE analysis the stresses of the soil skeleton at a given time instant t_{n+1} are computed, provided the equilibrium state at time instant t_n is known and, in addition, for the current time step Δt, estimates of the increments of the strain of the soil skeleton, $\Delta\varepsilon$, and of the capillary pressure and the air pressure, Δp^c and Δp^a, respectively are known. The plastic strain, net stress (or, alternatively, generalized effective stress), and hardening variable are then computed at t_{n+1}.

To this end, different stress update algorithms are available. In addition to robustness, the accuracy and efficiency of the employed stress update algorithm play a decisive role for FE analyses of geotechnical problems involving partially saturated soils. This motivated us for investigating different stress update algorithms for the Barcelona Basic Model, in particular

(a) an explicit stress update algorithm with adaptive sub-stepping and error control based on the Richardson extrapolation method;

(b) a general return mapping algorithm [4], which is characterized by backward Euler integration of the rate equations for the plastic strains and the hardening variable, and enforcing the yield condition $f = 0$ at t_{n+1}; it requires solving a system of nonlinear equations for the consistency parameter, the net stress, and the hardening variable;

(c) an optimized return mapping algorithm, which requires only solving a scalar non-linear equation for the first invariant of net stress;

(d) a semi-explicit stress update algorithm, which is characterized by explicit integration of the rate equations for the plastic strains and the hardening variable and by enforcing the yield condition $f = 0$ at t_{n+1} for determining the consistency parameter; it is combined with sub-stepping and error control along the lines of the explicit algorithm;

(e) an implicit 5th–order Runge–Kutta stress update algorithm with error control, proposed in [5].

The comparison of the investigated stress update algorithms is performed on the basis of two different sets of material parameters for the BBM, provided in [2]. For larger strain increments, it revealed relatively large integration errors of the return mapping algorithms. Hence, the latter are also enhanced with adaptive sub-stepping and error control.

A comparison of the efficiency of the investigated stress update algorithms for the prescribed maximum values of the integration error, ranging from 10^{-1} to 10^{-10}, is shown in Fig. 1. The diagrams are based on stress updates for 25 combinations of volumetric and deviatoric strain increments of 0.5%, 0.75%, 1.0%, 1.25%, and 1.5% and taking the mean values of the computed error and the mean values of the number of required arithmetic operations.

It follows from Fig. 1 that for a prescribed error threshold value, the optimized return mapping algorithm is by far more efficient than the general return mapping algorithm,

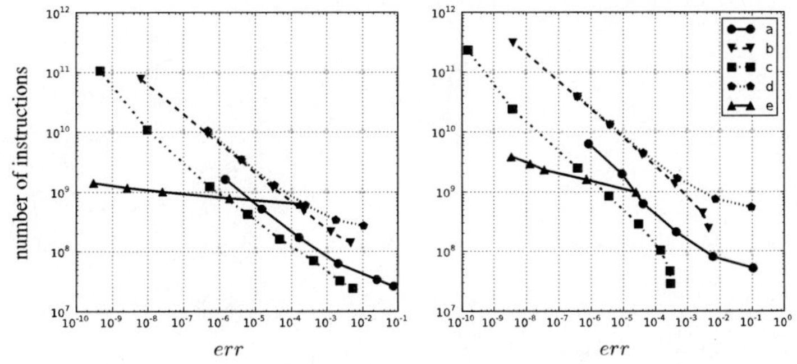

Fig. 1. Work precision diagrams for two sets of material parameters: (a) explicit stress update, (b) general return mapping algorithm, (c) optimized return mapping algorithm, (d) semi-explicit stress update algorithm, (e) implicit 5th–order Runge–Kutta algorithm

and it is even more efficient than the explicit integration method. The RADAU5 algorithm is very efficient for very small prescribed values of error tolerances.

3 APPLICATION

The developed three-phase model for partially saturated soils can be applied to the numerical simulation of a wide range of problems encountered in engineering practice. In the following, the application of the numerical model is demonstrated by the simplified 2D coupled solid–fluid FE analysis of water flow through a homogeneous earth dam (Fig. 2) due to reservoir impoundment and by the 3D coupled solid–fluid FE analysis of the injection of compressed air into water-saturated soil for displacing the groundwater. Figure 2 exemplarily shows the distribution of the capillary pressure for

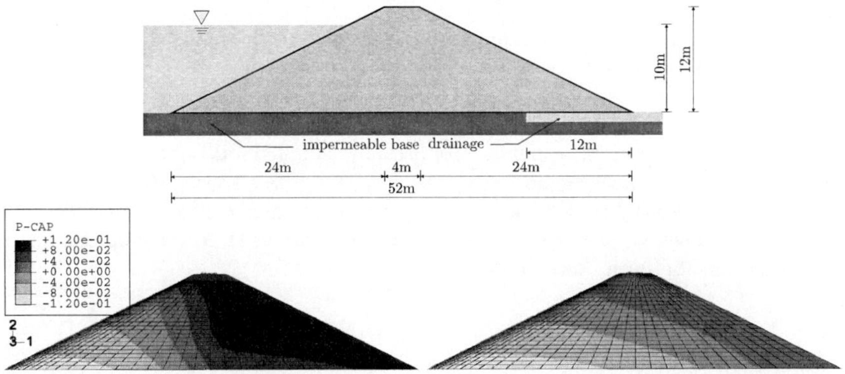

Fig. 2. Cross section of the homogeneous earth dam (top); distribution of the capillary pressure [MPa] (negative values denote water pressure in the water saturated domain) for selected time instants (bottom): t = 60 d (left), t = 800 d (right)

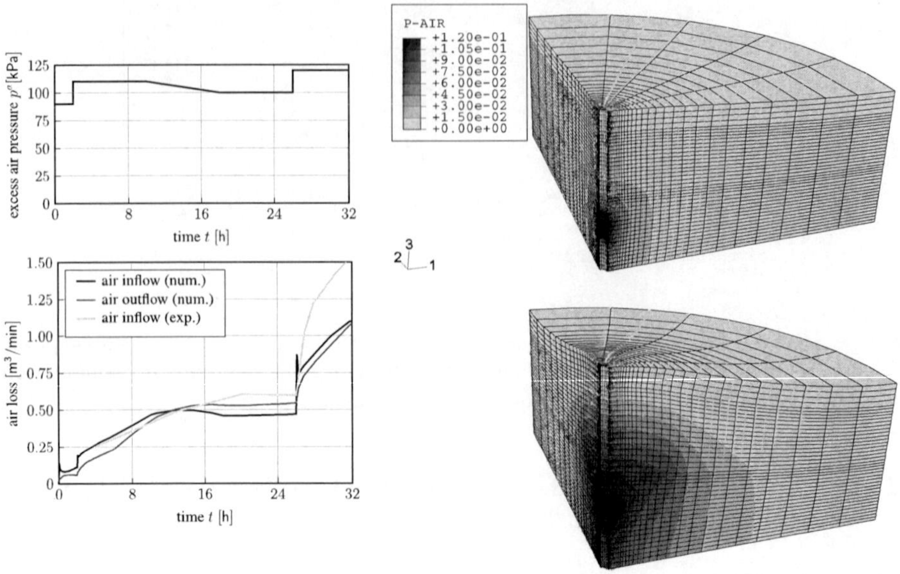

Fig. 3. Applied excess air pressure (top, left), measured and computed air inflow and outflow (bottom, left) and deformed FE mesh (displacement magnification factor: 1000) with distributions of the excess air pressure [MPa], 2 hours (top, right) and 32 hours (bottom, right) after beginning of the test

two selected time instants. Figure 3 contains selected results of the numerical simulation of injecting compressed air into water-saturated soil at a depth ranging from 13 m to 16 m below the surface by means of a bore hole with a diameter of 1.5 m.

REFERENCES

[1] Lewis, R.W., Schrefler, B.A.: The finite element method in the static and dynamic deformation and consolidation of porous media. John Wiley & Sons, Chichester (1998)

[2] Alonso, E.E., Gens, A., Josa, A.: A constitutive model for partially saturated soils. Géotechnique 40, 405–430 (1990)

[3] Kohler, R., Hofstetter, G.: A cap model for partially saturated soils. International Journal for Numerical and Analytical Methods in Geomechanics 32, 981–1004 (2008)

[4] Simo, J.C., Hughes, T.J.R.: Computational inelasticity. Springer, New York (1998)

[5] Hairer, E., Wanner, G.: Solving ordinary dierential equations II, stiff and differential- algebraic problems. Springer, Berlin (2002)

NUMERICAL MODELING OF POROMECHANICS IN CONTINUOUS AND LOCALIZED REGIONS

Craig D. Foster[1] and Talisa Mohammad Nejad[2]

[1] Department of Civil and Materials Engineering
University of Illinois at Chicago
Department of Civil and Materials Engineering (MC 246)
3085 Engineering Research Facility
842 West Taylor Street
Chicago, IL 60607-7023
e-mail: fosterc@uic.edu, web page: http://www.uic.edu/depts/cme/people/faculty/fosterc.html
[2] University of Illinois at Chicago
Department of Civil and Materials Engineering (MC 246)
2095 Engineering Research Facility
842 West Taylor Street
Chicago, IL 60607-7023
e-mail: tmoham5@uic.edu

Summary. We develop a model for porous flow, which includes both continuum and fractured flow regions using an enhanced strain finite element framework. Continuum flow is based on Darcy's law, with an additional term for the flow on the fracture, based the dilation on the crack. The porous flow model is embedded in a finite element framework, with the flow along a fracture handled using an enhanced element that contains an additional permeability component for fluid flow along the fracture. The enhanced permeability element will be coupled with the enhanced element for deformation along the fracture. Such elements exist, and the important quantities of pore pressure and volumetric deformation, both in bulk material and along the fracture, are passed between the two problems.

Keywords: fractured flow, poromechanics, porous flow.

1 INTRODUCTION

Fluid flow through porous media has many applications, including geotechnical, geophysical, and biological. In many of these situations, localized surfaces or narrow regions such as fractures, shear bands, or compaction bands play in important role both in the mechanics and fluid flow characteristics. Examples include zones of fissured rock, earthquake faults, and slope stability problems.

While there are relatively good physical models of how mechanical and fluid flow processes interact at the continuum scale, there are complex interactions along localized zones that also interact with the bulk processes. Permeability characteristics may change dramatically both along the localized band or across it. Mechanically, the changing fluid pressures may reduce the effective stress along the localized zone, which may already be weak, allowing for further deformation in that region. Further deformation

R.I. Borja (Ed.): Multiscale and Multiphysics Processes in Geomechanics, SSGG, pp. 141–144.
springerlink.com

on the band can result in dilation or compaction, which changes both the volume of the fluid in the localized region and the permeability characteristics. Changes in the stress field could also result in propagation of the localized zone.

We develop a model for porous flow processes at both continuum and localized regions using an enhanced strain finite element framework. Enhanced strain elements allow for the insertion of failure surface through the element at a critical orientation and location within the element, and have been used extensively for modeling propagating discontinuities.

The enhanced element will be coupled with a multiphysics approach. While finite element models exist for coupled solid deformation and fluid flow, fluid transfer between bulk material localized zones has only been explored by a few researchers, e.g. [4, 5]. Fluid pressure affects slip in the localized zone via an effective stress, while increased permeability in the band increases fluid flow in an anisotropic manner.

2 PERMEABILITY MODEL

Bulk permeability is modeled using Darcy's law $\boldsymbol{v} = -\boldsymbol{k}\boldsymbol{\nabla} h$, where $\boldsymbol{v}$ is the superficial velocity, $\boldsymbol{k}$ is the bulk permeability matrix, and h is the total head. In addition, there is fluid flow along the fracture given by

$$\hat{\boldsymbol{v}} = -\hat{k}(A_S)\hat{\boldsymbol{i}} \tag{1}$$

where $\hat{\boldsymbol{v}}$ is the superficial velocity along the crack, $\hat{k}$ is the permeability along the crack, which varies with the cross-sectional area A_S, and $\hat{\boldsymbol{i}}$ is gradient of the head within the fracture. The gradient can be be expressed in a finite element framework as

$$\hat{\boldsymbol{i}} = \frac{h_A - h_B}{L_S} \frac{\boldsymbol{x}_A - \boldsymbol{x}_B}{L_S} \tag{2}$$

where h_A and h_B are respectively, the total head values at the end nodes of the crack segment, $\boldsymbol{x}_A$ and $\boldsymbol{x}_B$ are the positions of the nodes, and L_S is the length of the crack. Hence, the first fraction determines the magnitude of the gradient, and the second the direction.

3 FINITE ELEMENT IMPLEMENTATION

The finite element model solves for the total head h as a primary variable. While enhanced finite elements for solid deformation condense the added degrees of freedom associated with the localized surface at the element level, here we add the degrees of freedom at the intersection of the surface with each element edge, to determine a hydraulic gradient along the band. To minimize the direct interaction of the flow on either side of the band, the element is "cut" by the band and divided into sub-elements over which the necessary quantities are integrated separately (see Figure 1). The additional nodal values are added at the end of the the respective vectors. Elements at end of the band can be accommodated using extra nodes only on the side where the band meets the element. Piecewise linear shape functions may then be used on that side, using standard transition element principles.

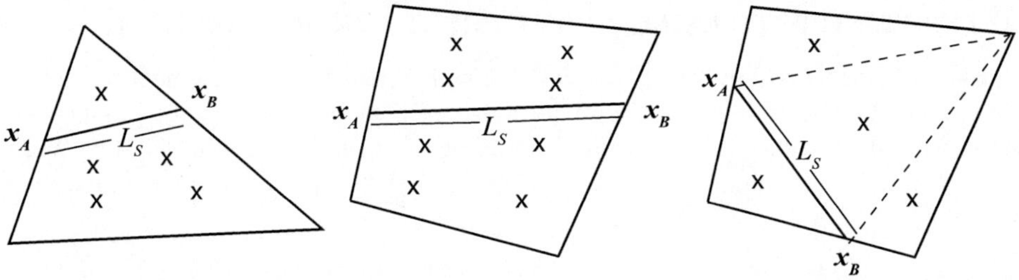

Fig. 1. Examples of enhanced elements with linear edges. Enhanced elements are divided by the fracture surface, and the two sides of the element are integrated separately. An additional permeability term connects the two nodes at the end of the fracture surface.

Standard development of the Galerkin form leads to a modified element internal "force" vector of

$$\boldsymbol{f}^e_{int} = \int_\Omega \boldsymbol{B}^{eT} \boldsymbol{v} \mathrm{d}\Omega \tag{3}$$

$$= \int_\Omega \boldsymbol{B}^{eT} \left\{ \boldsymbol{k} \nabla h + \frac{\hat{k}}{L_S^2} \left[(\boldsymbol{x}_A - \boldsymbol{x}_B) \otimes \boldsymbol{Q}^e \right] \boldsymbol{h}^e \delta_S \right\} \mathrm{d}\Omega \tag{4}$$

$$= \int_\Omega \boldsymbol{B}^{eT} \left\{ \boldsymbol{k} \boldsymbol{B}^e + \frac{\hat{k}}{L_S^2} \left[(\boldsymbol{x}_A - \boldsymbol{x}_B) \otimes \boldsymbol{Q}^e \right] \delta_S \right\} \mathrm{d}\Omega \boldsymbol{h}^e \tag{5}$$

$$= \left\{ \int_\Omega \boldsymbol{B}^{eT} \boldsymbol{k} \boldsymbol{B}^e \mathrm{d}\Omega + \int_S \boldsymbol{B}^{eT} \frac{\hat{k}}{L_S^2} \left[(\boldsymbol{x}_A - \boldsymbol{x}_B) \otimes \boldsymbol{Q}^e \right] \mathrm{d}S \right\} \boldsymbol{h}^e. \tag{6}$$

The stiffness matrix may then be written as

$$\boldsymbol{k}^e = \frac{\mathrm{d}\boldsymbol{f}^e_{int}}{\mathrm{d}\boldsymbol{h}^e} = \int_\Omega \boldsymbol{B}^{eT} \boldsymbol{k} \boldsymbol{B}^e \mathrm{d}\Omega + \int_S \boldsymbol{B}^{eT} \frac{\hat{k}}{L_S^2} \left[(\boldsymbol{x}_A - \boldsymbol{x}_B) \otimes \boldsymbol{Q}^e \right] \mathrm{d}S \tag{7}$$

where

$$\boldsymbol{Q}^e = \left\{ \begin{array}{c} 0 \\ . \\ . \\ . \\ 0 \\ 1 \\ -1 \end{array} \right\} \begin{array}{l} \\ \\ \\ \\ \\ \leftarrow \texttt{position A} \\ \leftarrow \texttt{position B} \end{array} \tag{8}$$

The second term arises from the flow on the fractured surface. With these enhancements, element matrices can assembled and the problem solved using standard techniques.

4 COUPLING WITH ENHANCED ELEMENT FOR DEFORMATION

These elements will eventually be coupled with enhanced strain element for deformation [1-3]. The coupling will initially solve each problem separately, passing the necessary information – pore pressure, volumetric strain, and crack dilation — between the two codes. The solution for each time step is thereby reached through an iterative process.

5 CONCLUSIONS

In summary, we have develop an enhanced finite element for porous flow that includes bulk permeability and flow between fractured surfaces. The permeability on the fractured surfaces varies with dilation along the interface. This development allows the permeability model to be coupled with an enhanced element for deformation that accounts for localized deformation along the fracture.

REFERENCES

[1] Simo, J.C., Oliver, J., Armero, F.: An analysis of strong discontinuities induced by strain-softening in rate-independent inelastic solids. Comp. Mech. 12, 277–296 (1993)

[2] Borja, R.I., Regueiro, R.A.: Strain localization in frictional materials ex-hibiting displacement jumps. Comp. Methods Appl. Mech. Eng. 37, 3333–3338 (2001)

[3] Foster, C.D., Borja, R.I., Regueiro, R.A.: Embedded strong discontinuity finite elements for fractured geomaterials with variable friction. Int. J. Num. Method Eng. 112, B04301(2007)

[4] Armero, F., Callari, C.: An analysis of strong discontinuities in a saturated poro-plastic solid. Int. J. Num. Method Eng. 46(10), 1673–1698 (1999)

[5] Rethore, J., Borst, R., Abellan, M.: A discrete model for the dynamic propagation of shear bands in a fluid-saturated medium. Int. J. Numer. Anal. Meth. Geomech. 31(2), 347–370 (2007)

EARTHQUAKE RUPTURES ON ROUGH FAULTS

Eric M. Dunham[1], Jeremy E. Kozdon[1], David Belanger[2], and Lin Cong[3]

[1] Department of Geophysics
Stanford University
397 Panama Mall
Stanford, CA 94305
e-mail: edunham@stanford.edu, jkozdon@stanford.edu
[2] BBN Technologies
10 Moulton St
Cambridge, MA 02138
e-mail: dbelange@bbn.com
[3] Stanford Graduate School of Business
518 Memorial Way
Stanford, CA 94305
e-mail: wtsung@stanford.edu

Summary. Natural fault surfaces exhibit roughness at all scales, with root-mean-square height fluctuations of order 10^{-3} to 10^{-2} times the profile length. We study earthquake rupture propagation on such faults, using strongly rate-weakening fault friction and off-fault plasticity. Inelastic deformation bounds stresses to reasonable values and prevents fault opening. Stress perturbations induced by slip on rough faults cause irregular rupture propagation and the production of incoherent high-frequency ground motion.

Keywords: earthquakes, faults, friction, roughness.

1 INTRODUCTION

While frequently idealized as infinitesimally thin planar surfaces, natural faults exhibit a variety of geometrical complexities. In this study, we focus on one ubiquitous type of complexity, fault roughness, and explore how deviations from planarity influence the earthquake rupture process. Roughness is observed at all scales using a variety of techniques (laser profilometers from $\sim$10 μm to $\sim$10 cm, ground-based LiDAR from $\sim$0.1 m to $\sim$100 m, and mapping of surface traces at the largest scales) [1, 2, 3]. Studies that combined data across the full range of scales suggest that fault surfaces are self-similar. Specifically, one-dimensional profiles along the surface (with zero mean), $y = h(x)$, are characterized by a power spectral density of the form $P_h(k) = (2\pi)^3\alpha^2k^{-3}$, for wavenumber k and amplitude-to-wavelength ratio α. The root-mean-square (rms) roughness between wavenumbers $k_{\min}$ and $k_{\max}$ is dominated by the largest wavelengths; for a profile of length L, which has roughness at all wavelengths less than L:

$$h_{rms}(k_{\min} = 2\pi/L, k_{\max} \to \infty) = \alpha L. \tag{1}$$

R.I. Borja (Ed.): Multiscale and Multiphysics Processes in Geomechanics, SSGG, pp. 145–148.
springerlink.com

It is thus evident that the rms deviations from planarity scale with the profile length. Alternatively, if the profile is decomposed into Fourier modes, the amplitude of each sinusoidal component increases with wavelength in a manner that satisfies equation (1).

Faults are smoother in the direction of slip ($\alpha \sim 10^{-3} - 10^{-2}$, with the smaller values being more appropriate for mature faults) than in the slip-perpendicular direction ($\alpha \sim 10^{-2}$). In this study, we focus solely on roughness in the direction of slip, and consider only roughness wavelengths that are much larger than the characteristic amount of slip in a single earthquake (an assumption required for our analytic and numeric analyses). An example of a synthetically generated self-similar fault is shown in Figure 1.

Slip on nonplanar faults perturbs the local stress field and increases the resistance to slip [4, 5]. Insight into these effects can be obtained using boundary perturbation techniques to develop approximate solutions to quasi-static linear elasticity problems. In contrast to the rms roughness, the rms stress perturbations are dominated by the shortest wavelengths of roughness. For slip of magnitude Δ across a fault with constant friction coefficient, the rms normal stress perturbation is [6]

$$\sigma_{rms}(k_{\min} \to 0, k_{\max} = 2\pi/\lambda_{\min}) = 2\pi^2 \alpha \frac{G}{1-\nu} \frac{\Delta}{\lambda_{\min}}, \tag{2}$$

where G is shear modulus, and ν is Poisson's ratio. For reasonable parameters, the normal stress perturbation predicted by this analysis will exceed the background compressive effective stress levels on faults at seismogenic depths, implying that the walls of the fault should open. That seems unlikely, as such large stress perturbations will almost certainly be prevented by inelastic deformation within damage zones surrounding the fault.

2 NUMERICAL MODEL

To investigate the effects of roughness in more detail, we model rupture propagation on nonplanar faults in an infinite, homogeneous medium under conditions of plane strain. The fault is a synthetically generated band-limited self-similar profile, as shown

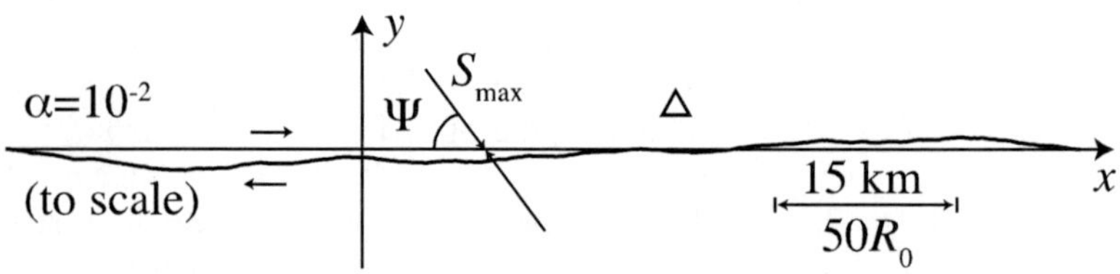

Fig. 1. Band-limited self-similar fault profile, $y = h(x)$, with amplitude-to-wavelength ratio $\alpha = 10^{-2}$. Synthetic seismograms are calculated at station marked with triangle in (a). The maximum principal prestress, $S_{\max}$, is inclined at angle Ψ to $y = 0$. Fault strength drops over a distance of $\sim R_0$ (a length scale emerging from friction and elasticity); $R_0 = 300$ m is used in dimensional scales. Roughness is present at wavelengths $\lambda \geq \lambda_{\min} = 1.25 R_0$ ($= 375$ m) in this example.

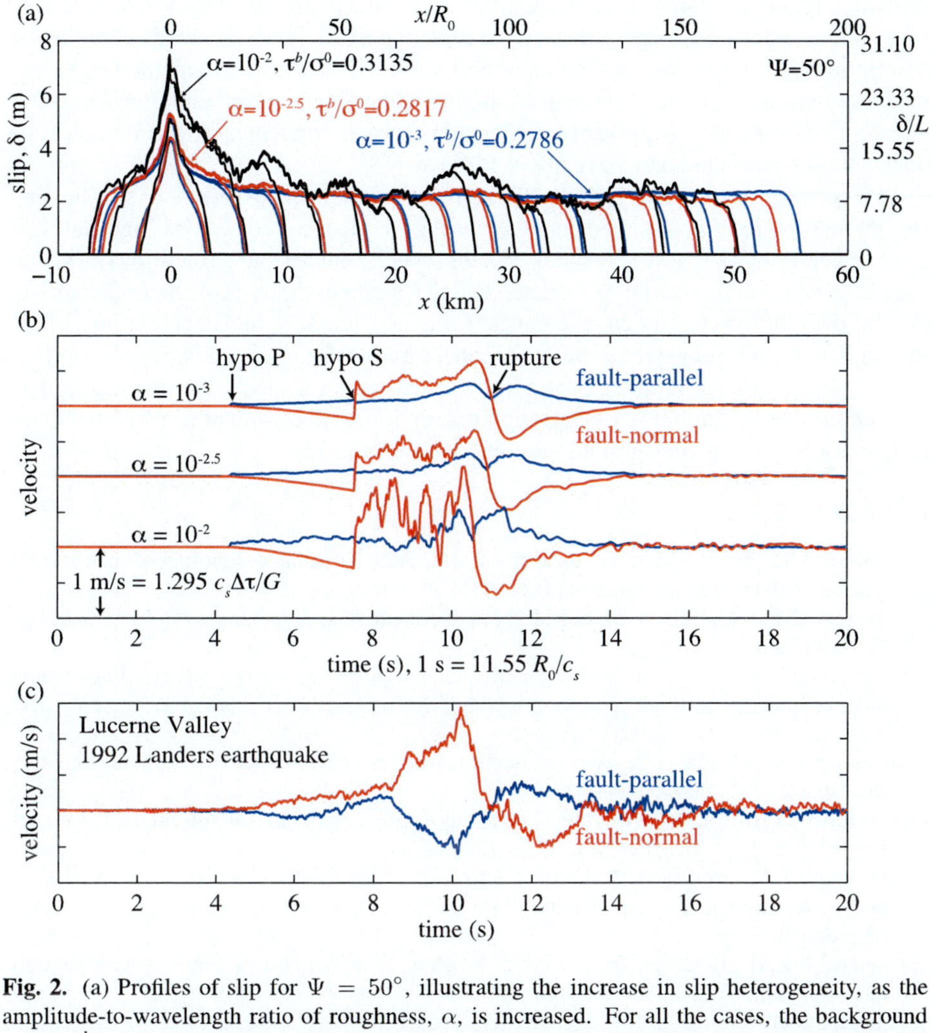

Fig. 2. (a) Profiles of slip for $\Psi = 50^\circ$, illustrating the increase in slip heterogeneity, as the amplitude-to-wavelength ratio of roughness, α, is increased. For all the cases, the background stress, τ^b, is just slightly above that required for self-sustaining propagation; while this value of τ^b increases with α, the average amount of slip remains roughly constant. (b) Synthetic velocity seismograms at station shown in Figure 1 for several values of α. Hypocentral P- and S-wave arrivals are marked; the two-sided fault-normal pulse occurs when the rupture passes the station. (c) The Lucerne Valley record from the 1992 M_w 7.3 Landers earthquake (2 km from the fault).

in Figure 1, with a lower wavelength cutoff introduced to ensure numerical resolution of all roughness features. The medium is modeled as a Drucker–Prager elastic–viscoplastic solid without cohesion. Plasticity prevents the development of unreasonably large stress perturbations, and, at least for the specific model employed in this study, completely eliminates fault opening. The fault is governed by a rate-and-state friction law featuring a strongly velocity-weakening steady-state response. (For further description, see [7, 6].) We use a summation-by-parts finite difference method on block-structured meshes; irregular geometries are handled using a coordinate transformation technique. The method is provably stable and accurate (convergence tests demonstrate that solutions presented here have about 1% error) [8].

One consequence of strongly rate-weakening friction is the existence of self-healing slip pulse ruptures for a limited range of background stress conditions (and multiple lines of reasoning suggest that natural faults operate under such conditions); we focus exclusively on this part of parameter space. Roughness introduces heterogeneity in the slip distribution (Figure 2a) and causes rapid accelerations and decelerations of the rupture front; both processes generate incoherent high-frequency seismic waves (Figure 2b). Synthetic seismograms from our simulations bear close resemblance to data from actual events (Figure 2c), suggesting that fault roughness might be responsible for incoherent high-frequency ground motion from earthquakes.

REFERENCES

[1] Power, W.L., Tullis, T.E.: Euclidean and fractal models for the description of rock surface roughness. J. Geophys. Res. 96, 415–424 (1991)

[2] Sagy, A., Brodsky, E.E., Axen, G.J.: Evolution of fault-surface roughness with slip. Geol. 35, 283–286 (2007)

[3] Candela, T., et al.: Characterization of fault roughness at various scales: Implications of three-dimensional high resolution topography measurements. Journal Pure Appl. Geophys 166, 1817–1851 (2009)

[4] Chester, F.M., Chester, J.S.: Stress and deformation along wavy frictional faults. J. Geophys. Res. 105, 23,421–23,430 (2000)

[5] Dieterich, J.H., Smith, D.E.: Nonplanar faults: Mechanics of slip and off-fault damage. Journal Pure Appl. Geophys. 166, 1799–1815 (2009)

[6] Dunham, E.M., Belanger, D., Cong, L., Kozdon, J.E.: Earthquake ruptures with strongly rate-weakening friction and off-fault plasticity: 2. Nonplanar faults. Bull. Seism. Soc. Am. (in press, 2011)

[7] Dunham, E.M., Belanger, D., Cong, L., Kozdon, J.E.: Earthquake ruptures with strongly rate-weakening friction and off-fault plasticity: 1. Planar faults. Bull. Seism. Soc. Am.(in press, 2011)

[8] Kozdon, J.E., Dunham, E.M., Nordström, J.: Interaction of waves with frictional interfaces using summation-by-parts difference operators: Weak enforcement of nonlinear boundary conditions. J. Sci. Comput. (submitted)

EARTHQUAKE SEQUENCE CALCULATIONS WITH DYNAMIC WEAKENING MECHANISMS

Hiroyuki Noda[1,*], Nadia Lapusta[1,2], and James R. Rice[3]

[1] Division of Geological and Planetary Sciences
California Institute of Technology
1200 E. California Blvd. Pasadena CA 91125
* e-mail: hnoda@caltech.edu
[2] Division of Engineering and Applied Science
California Institute of Technology
1200 E. California Blvd. Pasadena CA 91125
e-mail: lapusta@its.caltech.edu, web page: http://www.gps.caltech.edu/~lapusta/
[3] Department of Earth and Planetary Sciences and School of Engineering and Applied Sciences
Harvard University
29 Oxford Street, Cambridge, MA 02138
e-mail: rice@seas.harvard.edu, web page: http://esag.harvard.edu/rice/

Summary. There are multiple lines of evidence that major faults which host large earthquakes (e.g., San Andreas fault) operate at much lower shear stresses than what is predicted by Byerlee's law and a litho- and hydro-static stress state. Recent laboratory experimental studies suggest that a fault weakens dramatically at coseismic slip rates. We have taken the experimentally derived dynamic weakening into account in the earthquake sequence simulations. Our studies reveal that the overall stress level at which a fault operates is controlled by the frictional resistance at coseismic slip rates.

Keywords: friction, frictional heating, sequence of earthquakes, strong rate-weakening.

1 INTRODUCTION

Mature faults sustain low shear stress interseismically and during coseismic slip, as indicated by measurements of near-fault stress orientations (e.g., [1]) and observations of low heat flux [2]. At the same time, earthquake stress drops are typically between 1 and 10 MPa (e.g., [3]) which is much smaller than an expected value of shear strength at seismogenic depths (~ 100 MPa). These observations suggest that most fault points have low shear stress before large earthquakes, slip mainly at even lower shear stresses, and lock with the final stress only modestly lower than the initial value. [4], [5], and [6] demonstrated that even if a fault is strong at low slip rates, a dynamic rupture can propagate on a low-stressed fault (shear stress/normal stress ~ 0.3 or less) if the strength of a fault dramatically decreases coseismically, and discussed the operation of a fault at a low long-term shear stress.

Our aim is to elucidate effects of coseismically activated weakening mechanisms on the long-term fault behavior by conducting earthquake sequence simulations, following [4]. We have expanded the methodology that allows to simulate long slip histories while accounting for inertial effects during earthquakes [7, 8] to include strong dynamic weakening due to flash heating (FH) of microscopic contacts and

R.I. Borja (Ed.): Multiscale and Multiphysics Processes in Geomechanics, SSGG, pp. 149–152.
springerlink.com

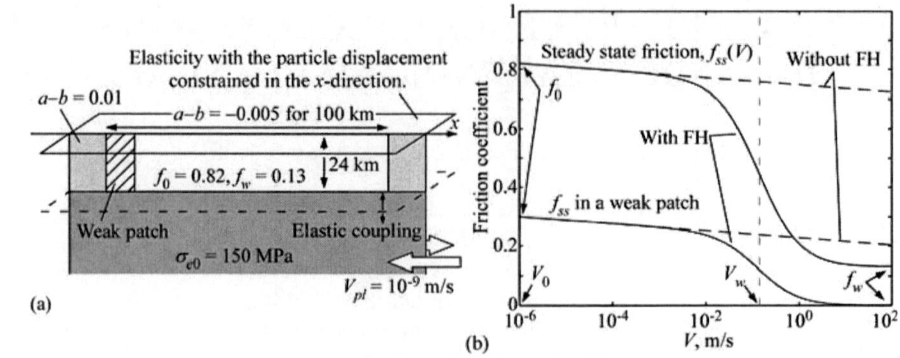

Fig. 1. (a) A schematic diagram of the crustal-plane model. (b) Steady-state friction coefficient, $f_{ss}(V)$, in the law with strong rate weakening due to FH

thermal pressurization (TP) of pore fluids. Improvements over [4] include a more stable numerical algorithm and code parallelization that allows exploring more realistic parameter regimes.

2 THERMAL WEAKENING MECHANISMS

Recent experimental studies (e.g. [9]) revealed that friction coefficients of rocks dramatically decrease at high slip rates (> 0.1 m/s). Such behavior can be explained by highly concentrated heat production in narrow shear zones and the resulting effects [10].

One such effect is temperature rise at microscopic frictional contacts (FH) which predicts $f \sim 1/V + const.$ at V larger than about 0.1 m/s, where f is the friction coefficient, and V is the slip rate. We add this effect to the Dieterich-Ruina's rate-and-state friction law (Figure 1b):

$$\tau = \sigma_e f(V,\theta) = \left(\sigma_n - p(y=0)\right) a \sinh^{-1}\left\{\frac{V}{2V_0}\exp\left(\frac{f_0 + b\ln\theta}{a}\right)\right\}$$

$$f_{ss}(V) = \frac{f(V, V_0/|V|) - f_w V/|V|}{1+|V|/V_w} + f_w \frac{V}{|V|} \equiv f(V,\theta_{ss}(V)), \quad \dot{\theta} = \frac{|V|}{L}\left(\theta_{ss}(V) - \theta\right)$$

where τ is the shear stress, σ_e and σ_n are respectively, the effective and total normal stresses, p is the pore pressure, y is the coordinate normal to the fault, θ is a state variable, a and b are respectively, rate and state parameters, L is the state evolution distance, V_0 and f_0 are the reference slip rate and friction coefficient, V_w is the slip rate at which FH becomes efficient, and f_w is the residual friction coefficient. Note that the aging effect in this law is important for simulating earthquake sequences including interseismic periods.

3 EARTHQUAKE SEQUENCES IN 2D CRUSTAL PLANE MODELS

In our simulation example, a 24-km-deep elastic plate (depth-averaged to a crustal plane) is loaded by steady-state slip at 10^{-9} m/s at the deeper fault extension (Figure 1a). The depth-averaged elastodynamic equation for the strike-parallel displacement $u(x, y, t)$ incorporates the plate loading V_{pl} and is given as eq. (A1) in [8]. On the fault, slip rate and shear stress are given by $V = 2(\partial u(x, y=0^+,t)/\partial t)$ and $\tau = \mu\times(\partial u(x, y=0^+,t)/\partial y)$, respectively. We examine the behavior of a 100-km rate-weakening region surrounded by stable rate-strengthening areas. The region has a 10-km weak region of low f_0 at one end, which produces frequent nucleation of seismic ruptures. $f_{ss}(V)$ in the seismogenic region is shown in Figure 1b.

Although the strength at each point on the fault at low slip rates is high ($\sigma_e f_0$ ~ 100 MPa), averaged shear stress on a fault is low because of low coseismic shear stress and frequent rupture nucleation, consistent with the results of [4], [5], [6]. Long-term

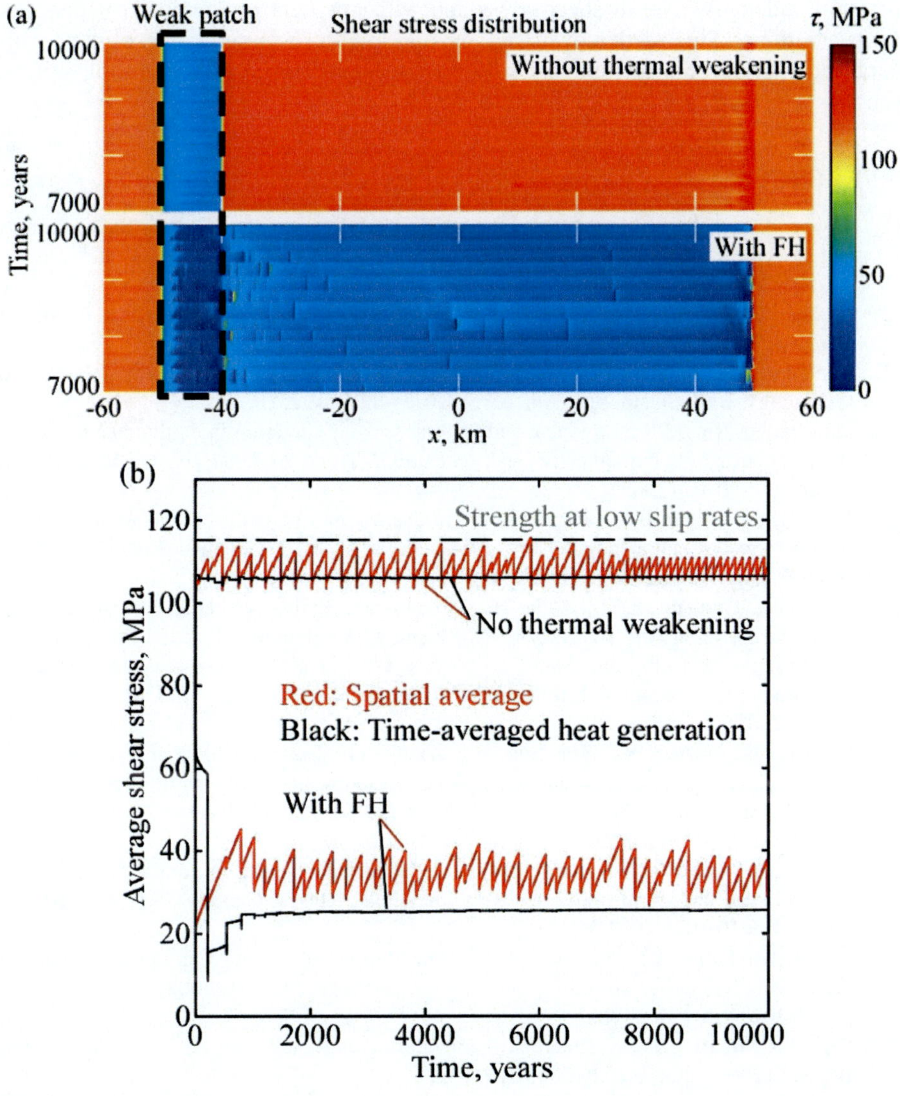

Fig. 2. (a) Shear stress distribution between 7000 and 10000 years from the initiation of calculation without thermal weakening (top) and with FH (bottom). (b) Average shear stress (red) and time-averaged heat generation rate (black) in the seismogenic region (-50 km $< x < 50$ km). Interseismic shear stress and long-term heat generation are significantly reduced by thermal weakening due to flash heating

shear stress distribution and averaged shear stress in the seismogenic region (-50 km < x < 50 km) are shown in Figures 2a and b, respectively, for cases with L = 10 mm.

4 CONCLUSIONS

Fault models that combine high static strength as shown in the lab, low dynamic strength due to flash heating, and locations for earthquake nucleation result in fault operation under low overall shear stress and with low heat production, as supported by observations. This is demonstrated by our long-term simulations of fault slip, which incorporate both tectonically slow loading and inertial effects during seismic events.

REFERENCES

[1] Townend, J., Zoback, M.D.: Regional tectonic stress near the San Andreas fault in central and southern California. Geophys. Res. Lett. 31, L15S11 (2004), doi:10.1029/203GL018918

[2] Lachenbruch, A.H., Sass, J.H.: Heat flow and energetics of the San Andreas fault zone. J. Geophys. Res. 85, 6185–6223 (1980)

[3] Kanamori, H., Anderson, D.L.: Theoretical basis of some empirical relations in seismology. Bull. Seism. Soc. Am. 65(5), 1073–1095 (1975)

[4] Lapusta, N., Rice, J.R.: Low-heat and low-stress fault operation in earthquake models of statically strong but dynamically weak faults. EOS Trans. AGU 84(46) Fall Meet. Suppl., Abstract S51B-02 (2003)

[5] Noda, H., Dunham, E.M., Rice, J.R.: Earthquake ruptures with thermal weakening and the operation of major faults at low overall stress levels. J. Geophys. Res. 114, B07302 27 (2009), doi:10.1029/2008JB006143

[6] Rice, J.R., Dunham, E.M., Noda, H.: Thermo- and hydro-mechanical processes along faults during rapid slip. In: Hatzor, Y., Sulem, J., Vardoulakis, I. (eds.) Proceedings of the Batsheva de Rothschild Seminar on Shear Physics at the Meso-scale in Earthquake and Landslide Mechanics, Ein Gedi, Israel, January 26-30, pp. 3–16. CRC Press, Boca Raton (2009)

[7] Lapusta, N., Rice, J.R., Ben-Zion, Y., Zheng, G.: Elastodynamic analysis for slow tectonic loading with spontaneous rupture episodes on faults with rate- and state-dependent friction. J. Geophys. Res. 105(B10), 23,765–23,790 (2000), doi:10.1029/2000JB900250

[8] Kaneko, Y., Lapusta, N.: Variability of earthquake nucleation in continuum models of rate-and-state faults and implications for aftershock rates. J. Geophys. Res. 113, B12312 (2008), doi:10.1029/2007JB005154

[9] Beeler, N., Tullis, T.E., Goldsby, D.L.: Constitutive relationships and physical basis of fault strength due to flash heating. J. Geophys. Res. 113, B01401 (2008), doi:10.1029/2007JB004988

[10] Rice, J.R.: Heating and weakening of faults during earthquake slip. J. Geophys. Res. 111, B05311 (2006), doi:10.1029/2005JB004006

FAULT LUBRICATION AND EARTHQUAKE PROPAGATION IN CARBONATE ROCKS

Nicola De Paola[1,*], Takehiro Hirose[2], Thomas Mitchell[3], Giulio Di Toro[4], Cecilia Viti[5], and Toshiko Shimamoto[3]

[1] Rock Mechanics Laboratory, Earth Sciences Department,
University of Durham, UK
*e-mail: nicola.de-paola@durham.ac.uk
[2] Kochi Institute for Core Sample Research,
Japan Agency for Marine-Earth Science and Technology, JAMSTEC
e-mail: hiroset@jamstec.go.jp
[3] Department of Earth and Planetary Systems Science,
Graduate School of Science, Hiroshima University, Japan
e-mail: tom-mitchell@hiroshima-u.ac.jp
[4] Istituto Nazionale di Geofisica e Vulcanologia, Italy
e-mail: giulio.ditoro@unipd.it
[5] Dipartimento di Scienze della Terra,
Universita' degli Studi di Siena, Italy
e-mail: viti@unisi.it

Summary. Friction experiments performed on dolomite/calcite at conditions typical during the propagation of large earthquakes (slip velocities $v > 1$ m/s and displacements $d > 1$ m) show that the frictional strength of experimental faults decays exponentially from peak values, in the Byerlees' range ($\mu_p \approx 0.8$), to extremely low steady-state values ($\mu_{ss} \approx 0.1$), attained over a weakening distance D_w. The integration of CO_2-emission data, recorded during the experiments, with microstructural observations shows that nanoparticles were produced in the slip zone due to cataclastic and thermally activated chemical/thermal processes (e.g. decarbonation reactions). Steady-state shear strength, soon after the onset of CO_2 emissions during the transient stage, is reduced below best fit exponential law. During the transient stage of dynamic weakening, flash heating temperature rises, greater than the value necessary to activate the thermal decomposition of dolomite (T=550°C), have been locally reached at the highly stressed frictional microcontacts. Pressurized fluids (CO_2), temporarily trapped within the slip zone, may also have contributed to dynamic weakening in accordance with the effective normal stress principle. Flash heating and thermal pressurization processes will be inhibited at the end of the transient stage as nanoparticles are produced and fluids can escape the slip zone. Steady-state dynamic weakening may be controlled by velocity-dependent frictional properties of nanoparticles. Seismic source parameters (e.g. slip weakening distance D_w and breakdown work W_b), calculated from experimental data, match those obtained by modelling of seismological data from earthquakes nucleated in the same carbonate rocks in Italy during the 1997 M=6 Colfiorito earthquakes. W_b scales with D_w according to a best fit power law, in a similar fashion to that inferred from existing seismological data sets.

Keywords: dynamic weakening, earthquake, flash heating, friction, thermal pressurization.

R.I. Borja (Ed.): Multiscale and Multiphysics Processes in Geomechanics, SSGG, pp. 153–156.
springerlink.com

1 INTRODUCTION

Several thermally activated slip weakening mechanisms have been suggested in the literature to account for low frictional strength of faults during earthquake propagation: flash heating and thermal pressurization [1], frictional melting [2], gel formation [3], thermal decomposition [4,5]. Growing field evidence from springs located nearby large active faults of CO_2 flux of crustal origin after large earthquakes e.g. Cofiorito epicentral area [6], led to suggest that earthquake propagation in carbonates, which are common seismogenic sources in the upper crust, may be favoured by dramatic fault weakening because of thermal processes (decarbonation reactions) that are activated when the fault rocks are driven into their instability field by frictional heating. In order to investigate the dynamic weakening processes which operate during earthquake propagation in thermally unstable rocks, a set of 34 friction experiments has been performed on dolomite and Mg-calcite granular rocks in a high-velocity friction apparatus at conditions that are typical during the propagation of large earthquakes (slip velocities v>1 m/s and displacements d>1 m). Mechanical and seismic source parameters obtained and calculated from experimental data have been compared with those obtained by modelling of seismological data from earthquakes nucleated in carbonate rocks in Italy during the 1997 M=6 Colfiorito earthquakes.

2 EXPERIMENTAL RESULT AND MICROSTRUCTURAL OBSERVATIONS

Experiments were performed on fine-grained (≈100 μm), sharp-edged, gouges of (a) pure dolomite $MgCa(CO_3)_2$ (dry and saturated with distilled water), (b) partially decomposed dolomite made of Mg-rich calcite $(Ca_xMg_{1-x})CO_3$ and periclase (MgO) and (c) totally decomposed dolomite made of lime (CaO) and periclase (MgO).

The experiments were run at normal stresses ranging from 0.4 to 2 MPa, slip rates ranging from 0.009 to 1.3 m/s, and displacements of 0.20–58 m, while monitoring CO_2-emission rates. For these normal stresses, and slip velocities $v > 0.69$ m/s, all gouge materials display a dramatic reduction in shear strength τ_f from initial peak values τ_p to steady-state values τ_{ss}, over a slip weakening distance D_w (Fig. 1a).

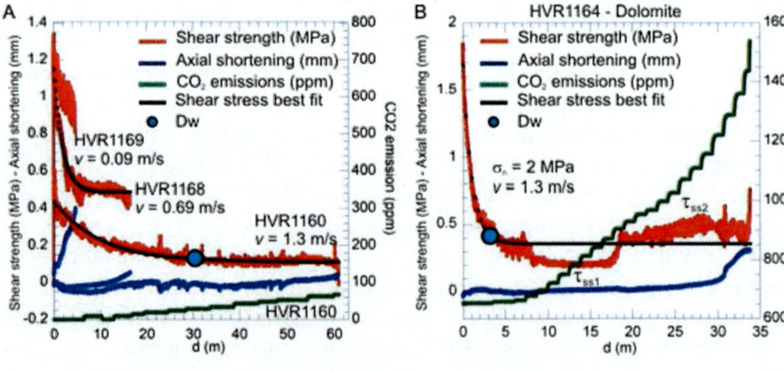

Fig. 1. (a) For v ≥ 0.69 m/s, τ_f is fitted by the exponential decay law $\tau = \tau_{ss} + (\tau_p - \tau_{ss}) e^{[\ln (0.05) d/Dw]}$ from peak (τ_p) to steady-state (τ_{ss}) values when plotted versus displacement. CO_2 emissions recorded during the experiment are small. (b): τ_{ss} is reduced (two τ_{ss} are attained) below best fit law following CO_2 emissions at the end of the transient stage.

A sharp increase in dynamic shear strength is observed for displacements larger than D_w, resulting in the attainment of a second steady-state $\tau_{ss2} > \tau_{ss1}$ (Fig. 1b). We interpret the increase in dynamic shear strength $\tau_{ss2} = \mu_{ss}\sigma'_n$ observed for d >> D_w, as a result of the increase of the effective normal stress $\sigma'_n = \sigma_n - P_f$, when pressurized CO_2 escapes from the localized slip zone and P_f is reduced.

All laboratory-deformed samples consistently show the localization of slip at the gouge-host rock boundary, within thin slip zones (50–100 μm) formed by fine- to ultra-fine-grained material. TEM scale observations show that the slip zone of the experimental samples is formed by a complex association of grains with relatively constant size, ranging from few nanometers up to 20 nm. Nano-grains in the slip zone are closely associated, giving rise to a compact, pore-free texture, with recurrent polygonalized boundaries. High-resolution images show that nano-grains are formed by ordered, defect-free sequences of lattice fringes, suggesting significant structural reorganization.

3 DISCUSSION AND CONCLUSIONS

3.1 Dynamic Weakening Mechanisms

The average calculated bulk temperature at the end of the transient stage is very close to T = 550 °C, necessary to initiate the thermal decomposition of dolomite. However, CO_2- emission data have been recorded before the end of the transient stage (Fig. 1), suggesting that flash heating temperature rise [1], greater than the value necessary to activate the thermal decomposition of dolomite (T= 550°C), have been locally reached at the highly stressed frictional microcontacts. The (lower) steady-state dynamic shear strength values $\tau_{ss1} = \mu_{ss}(\sigma_n - P_{f1})$, attained in the presence of pressurized fluids trapped within the slip zone (Fig. 1b), show little dependence on the normal stress (Fig. 2a). On the contrary, the (higher) steady-state dynamic shear strength values $\tau_{ss2} = \mu_{ss}(\sigma_n - P_{f2})$, attained when fluids could escape the slip

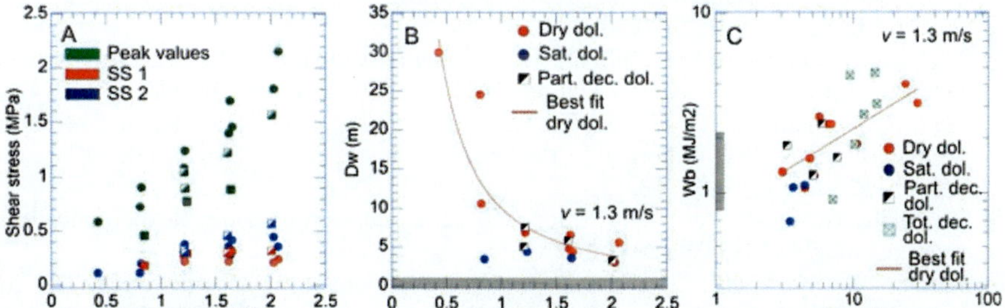

Fig. 2. (a) The diagram shows peak and steady-state stress $\tau_{ss2} > \tau_{ss1}$ for each material. (b): D_w decreases with increasing σ_n according to a power law least squares best fit equation $D_w = a\sigma_n^b$. (c): Breakdown work Wb, calculated from experimental data as the area of the surface below the dynamic shear strength curve increases with increasing displacement (D_w) according to a power law least squares best fit equation $W_b = aD_w^b$. The grey boxes in Figures 2b-c represent the ranges of values calculated by modelling of seismological data from the 1997 Colfiorito earthquakes [7].

zone ($P_{f1} > P_{f2}$) (Fig. 1b), increase with increasing σ_n (Fig. 2a). This behaviour is in agreement with the effective normal stress principle ($\sigma'_n = \sigma_n - P_f$).

During the transient stage of frictional strength decay from peak to steady-state values, dynamic weakening was mostly controlled by flash heating and thermal pressurization processes. However, flash heating would be largely inhibited at the end of the transient stage and during steady-state 1, when average grain size of particles in the slip zone is $\phi = 10–100$ nm, as gathered from microstructural observations. Finally, during the attainment of the second steady state, τ_{ss2} (Fig. 1b), fluids escaped the slip zone, and thermal pressurization ceased to operate. At this point, dynamic weakening was solely controlled by the velocity-dependent frictional properties of nanoparticles [8].

3.2 Seismic Source Parameters: Experimental vs. Seismological Data

At seismic slip rates of 1.3 m/s, D_w decreases with increasing normal stress according to a power law least squares best fit equation (Fig. 2b). Experimental D_w values are slightly higher than those obtained from dynamic modelling of the seismological data of the Colfiorito earthquakes (Fig. 2b, [5, 7]), which nucleated in the same rocks tested in this study. When extrapolated at seismogenic depths (σ_n = 20–90 MPa), experimental D_w data are well within the range of seismological values (0.3-1 m) (Fig. 2b). The breakdown work W_b data calculated from the experiments are within the same order of magnitude of the range of values (0.8–2.28 MJ/m^2) estimated for the Colfiorito earthquakes [5, 7] and scale with D_w according to a power law best fit equation (Fig. 2c). More generally, experimentally derived W_b and its scaling relationships with D_w are of the same order of magnitude and similar nature, respectively, when compared to the same parameters obtained from seismic inversion analyses of seismological data for well-studied moderate-to-large earthquakes [1, 5, 7] (Figs. 2b-c).

REFERENCES

[1] Rice, J.R.: Heating and weakening of faults during earthquake slip. J. Geophys. Res.-Sol. Ea. 111 (2006), doi:Artn B05311, 10.1029/2005jb004006

[2] Hirose, T., Shimamoto, T.: Growth of molten zone as a mechanism of slip weakening of simulated faults in gabbro during frictional melting. J. Geophys. Res.-Sol. Ea. 110 (2005), doi:Artn B05202, 10.1029/2004jb003207

[3] Di Toro, G., Goldsby, D.L., Tullis, T.E.: Friction falls towards Zero in Quartz Rock as slip velocity approaches seismic rates. Nature 427, 436 (2004), doi:10.1038/Nature02249

[4] Han, R., Shimamoto, T., Hirose, T., Ree, J.H., Ando, J.: Ultralow friction of carbonate faults caused by thermal decomposition. Science 316, 878–881 (2007), doi:10.1126/science.1139763

[5] De Paola, N., et al.: Fault lubrication and earthquake propagation in thermally unstable rocks. Geology 39, 35–38 (2011)

[6] Italiano, F., Martinelli, G., Plescia, P.: CO2 degassing over seismic areas: The role of mechanochemical production at the study case of Central Apennines. Pure Appl. Geophys. 165, 75–94 (2008)

[7] Tinti, E., Spudich, P., Cocco, M.: Earthquake fracture energy inferred from kinematic rupture models on extended faults. J. Geophys. Res.-Sol. Ea. 110 (2005)

[8] Han, R., Hirose, T., Shimamoto, T.: Strong velocity weakening and powder lubrication of simulated carbonate faults at seismic slip rates. J. Geophys. Res.-Sol. Ea. 115 (2010), doi:Artn B03412, 10.1029/2008jb006136

EQUIVALENT CONTINUUM MODELING FOR WAVE PROPAGATION IN DISCONTINUOUS MEDIA

Oleg Vorobiev

Lawrence Livermore National Laboratory L-231, P.O. Box 808, Livermore, CA 94550, USA
e-mail: vorobiev1@llnl.gov

Summary. This study presents numeric simulations of nonlinear wave propagating through jointed rock masses. The simulations were performed using the Lagrangian hydrocode GEODYN-L with joints treated explicitly using an advanced contact algorithm.

Keywords: equivalent continuum, jointed rock, wave propagations.

1 INTRODUCTION

Modeling the mechanical response of discontinuous media presents many computational challenges emphasized by various researchers. Analytic methods developed earlier for jointed rock masses [1, 2] can only be applied to small deformations and do not account for nonlinear irreversible processes taking place during large deformations. There have been attempts to use both Discrete Element Methods (DEM) and Finite Element Methods (FEM) to determine the properties of the fractured rock mass [3–5]. Most of them were aimed at quasi-static material response. The aim of this study is to develop a methodology to build EC models for nonlinear wave propagation through heavily jointed rock media. In contrast to many existing EC models for jointed rock, the current model considers regimes where responses both joints and rock material are strongly nonlinear. A discrete method was developed and designed for Lagrangian FE/FD codes, where all discontinuities are coincident with the mesh lines. The main advantage of such approach is its ability to treat large sliding and separation of the contacting surfaces which take place during large deformations of the representative volume.

2 MODELING APPROACH

2.1 Model for Intact Samples

Because of their heterogeneous nature, rock materials exhibit complicated mechanical behavior. They are brittle at low confinements and ductile at high pressures. To model this behavior, advanced plasticity models describing porous compaction and dilation, softening and hardening, and rate and scale dependence have been developed. Such models can be calibrated by comparing results of triaxial loading of a single material point with the experiments. The model used in this study operates in the frames of isotropic plasticity theory and was described in detail in [5].

R.I. Borja (Ed.): Multiscale and Multiphysics Processes in Geomechanics, SSGG, pp. 157–160.
springerlink.com

2.2 Model for Joints

It is known from experimental observations that joint normal closure is a nonlinear function of applied normal stress, resembling a hyperbola [8]. To describe the experiments, a nonlinear normal modulus and a constant friction coefficent was used at the joints. The details of the numerical implementation can be found in [6].

2.3 Numerical Method

Explicit finite-difference massively parallel code GEODYN-L [6, 7] was used in calculations. The joints between the blocks of rock were modeled using the Simple Common Plane contact algorithm described in [6]. The method used in [5] for numeric simulations of quasi-static loading of randomly jointed rock was also used in this study to generate randomly distributed joints. In this method, the numeric mesh covering the region was split into multiple meshes for the blocks of elements grouped by proximity to randomly seeded centers of spheres.

3 RESULTS

The study revealed that the response of jointed medium is strongly nonlinear and depends both on both material and joint properties. When the joints have a preferred orientation, the response is anisotropic, and when the joints are distributed randomly, the response is isotropic. Figure 1 shows that the UCS found in calculations for a cubic volume with three joint sets depends on the loading direction. The anisotropy is higher for the joints with low friction. The joint aperture increase reduces the anisotropy but makes the material weaker. Results of these simulations indicate that one need an anisotropic model to reproduce the response of the jointed rock with three orthogonal sets of joints.

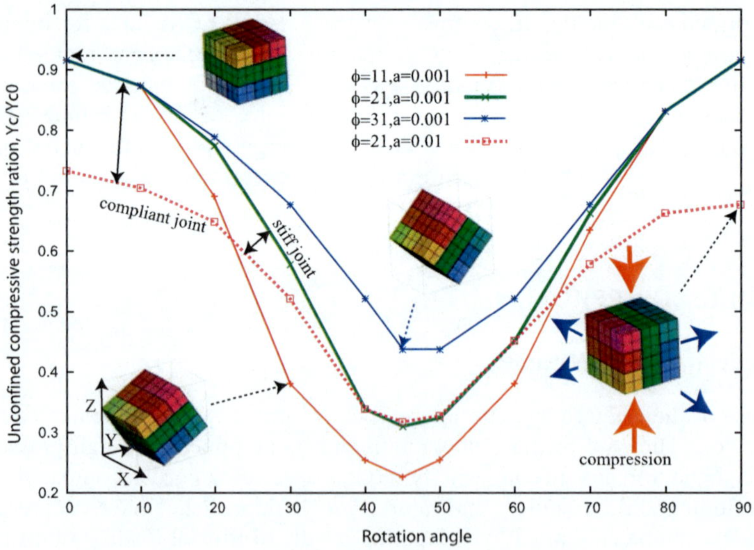

Fig. 1. Unconfined Compressive Strength (UCS) for Representative Volume (RV) with three orthogonal sets of joints for various friction angles and apertures

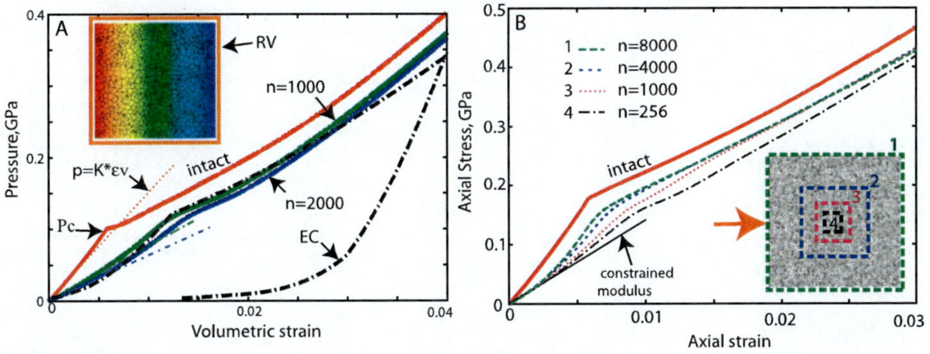

Fig. 2. A:-Pressure vs volumetric strain for quasi-static uniaxial loading (joint density effect), B:– Stress vs Strain for quasi-static uniaxial loading (RV size effect)

A natural way of building an equivalent continuum model for a randomly jointed media by loading a RV quasi-statically and fitting a homogeneous model by matching its response was considered [5]. Figure 2 shows that the initial bulk modulus is reduced with the increase of the number of joints. The dotted lines show initial bulk modulus calculated using analytic expressions [1]. As the medium is compacted, the joints become stiffer and, eventually, many more deformations are taken by the rock blocks. The transition from the quasi-elastic regime to the plastic compaction is not as distinct as in the case of solid material where the pores begin compacting at a fixed pressure,P_c , shown in Fig. 2A. Yet, the same compaction model as the one used for the intact material can be applied to the jointed medium if its parameters are modified to account for the presence of more compliant joints. It was found that volumetric response of RV can show some sensitivity to the volume size as shown in Fig. 2B. This is because the joint aperture remains constant while the RV size changes. If the wave amplitude is small, so that the displacement is less than the joint aperture, then compressibility of the RV remains roughly the same.The ratio of the aperture to the characteristic volume size defines how important the contribution of the joints into effective compressibility is. For example, the biggest RV (curve-1 with n=8000) gives the compaction curve which is close to ones of the intact material.

A study of plane waves propagating though a jointed medium showed that the deviatoric stress is not constant behind the plastic front as it would be if any homogeneous model were used. This relaxation is related to the multiple elastic waves running between the joints and the boundaries, which shift the equilibrium between the volumetric and deviatoric compressions. The axial stress in discrete simulations stays constant as well as the velocity and is proportional to the velocity value with the coefficient of proportionality being the effective impedance of the system. It is seen from Fig. 3 that the impedance is reduced when the number of joints is increased. Also, the rate of the deviatoric stress relaxation seems to increase with increasing number of joints. It is seen from the picture that, unlike in quasi-static loading, in the dynamic loading, the deviatoric stress drops after reaching its peak because of the stress relaxation. The rate-dependent as well as the strain-softening features of the intact model were not exercised in these calculations, and the stress relaxation effect was still observed.

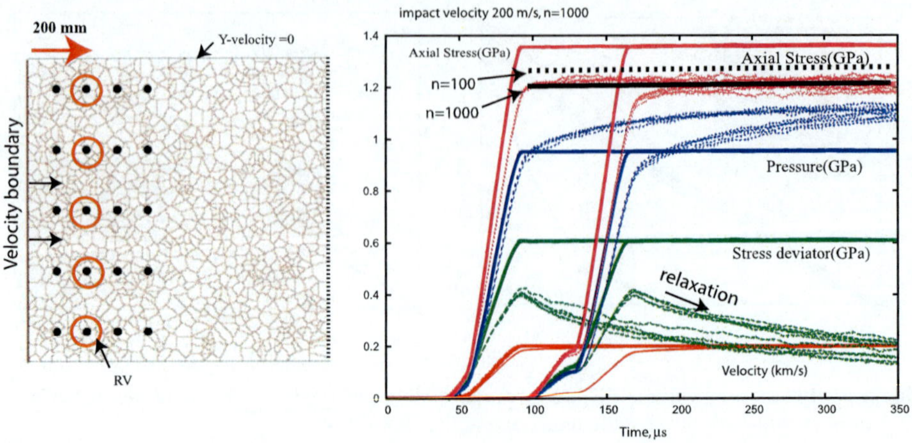

Fig. 3. Evolution of the stresses and velocity in RVs at two ranges(R=200,400) in a plane strain loading for 1000 blocks. The wave on the left was modeled using constant velocity boundary condition where the motion on the top and the bottom was constrained.

This study was performed under the auspices of the U.S. Department of Energy by Lawrence Livermore National Laboratory under Contract DE-AC52-07NA27344. This study was supported by the DTRA Basic Research Award IACROs 10-41581.

REFERENCES

[1] Fossum, A.: Effective elastic properties for a randomly jointed rock mass. Int. Journal Rock Mech. Min. Sciences Geomechanics Abstracts 22(6), 467–470 (1985)

[2] Cai, M., Horii, H.: A constitutive model of highly jointed rock masses. Mechanics of Materials 13, 217–246 (1992)

[3] Ki-Bok, M., Jing, L.: Numerical determination of the equivalent elastic compliance tensor for fractured rock masses using the distinct element method. Int. Journal Rock Mech & Mining Sciences 40, 795–816 (2003)

[4] Pouya, A., Ghoreychi, M.: Determination of rock mass strength properties by homogenization. Int. J. Num. Anal. Meth. in Geomechanics 25, 1285–1303 (2001)

[5] Vorobiev, O.: Generic strength model for dry jointed rock masses. Int. Journal of Plasticity 24(12), 2221–2247 (2008)

[6] Vorobiev, O.: Simple common plane contact algorithm. Comput. Meth. Appl. Mech. Engng (2010) (under revision)

[7] Vorobiev, O.: Discrete and continuum methods for numerical simulations of nonlinear wave propagation in discontinuous media. International Journal for Numerical Methods in Eng. 83, 482–507 (2010)

[8] Bandis, S.C., Lumsden, A.C., Barton, N.R.: Fundamentals of Rock Joint Deformation. Int. J. Rock. Mech. Min. Sci. Geomech. Abstr. 20(6), 249–268 (1983)

DYNAMIC FAULT RUPTURE AND PROPAGATION THROUGH A FIXED FINITE ELEMENT GRID

Fushen Liu[1] and Ronaldo I. Borja[2]

[1] Department of Civil and Environmental Engineering
Stanford University
Stanford, CA94305, USA
e-mail: fsliu@stanford.edu
[2] Department of Civil and Environmental Engineering
Stanford University
Stanford, CA94305, USA
e-mail: borja@stanford.edu

Summary. The extended finite element method is used to simulate dynamic fault rupture in elastoplastic solids. Rupture is governed by a linear slip-weakening law and finite element calculations are done explicitly. In a numeric example, we study a benchmark problem comparing the new technique with conventional finite element simulations, including the regular penalty formulation and split-node technique for rupture propagating with off-fault bulk plasticity.

Keywords: dynamic fault rupture, extended finite element method, off-fault plasticity, slip weakening.

1 INTRODUCTION

The finite element method has been widely used for simulating dynamic fault rupture processes, see [1, 2]. An explicit finite element scheme is typically combined with a split-node formulation and a linear slip-weakening law. Recently, the extended finite element method [3] has been used for modeling an arbitrarily evolving frictional discontinuity in solids. This method allows the propagation of displacement discontinuity through the interior of finite elements by introducing global enrichment to the finite element interpolations. In this article, we use an explicit version of the extended finite element method to study the dynamic fault rupture processes in elastoplastic solids. We enforce the frictional contact conditions on the fault surfaces by penalty method. A non-associated Mohr–Coulomb yield criterion is used to model the damage zone in the neighborhood of the fault. In a numeric example, we compare the performance of the proposed technique with traditional finite element methods that use the regular penalty formulation and split-node technique.

2 VARIATIONAL EQUATIONS

Following [4], a trial function $\boldsymbol{u}$ and a weighting function $\boldsymbol{\eta}$ are written as

$$\boldsymbol{u} = H_{\mathcal{S}}(\boldsymbol{x})\tilde{\boldsymbol{u}} + \bar{\boldsymbol{u}}, \tag{1}$$

R.I. Borja (Ed.): Multiscale and Multiphysics Processes in Geomechanics, SSGG, pp. 161–164.
springerlink.com

and

$$\eta = H_S(x)\tilde{\eta} + \bar{\eta}, \tag{2}$$

where H_S is a sign function, $\tilde{u}$ and $\tilde{\eta}$ are the discontinuous displacements, and $\bar{u}$ and $\bar{\eta}$ are the continuous displacements. We thus obtain two sets of independent variational equations:

$$\int_\Omega \nabla^s \bar{\eta} : \sigma \mathrm{d}\Omega = \int_\Omega \bar{\eta} \cdot (f - \rho \ddot{u}) \mathrm{d}\Omega + \int_\Gamma \bar{\eta} \cdot t \mathrm{d}\Gamma, \tag{3}$$

and

$$\int_\Omega [H_S(x) \nabla^s \tilde{\eta}] : \sigma \mathrm{d}\Omega + \mathcal{G}_c(\tilde{\eta}, t_{S_-}) = \int_\Omega H_S(x)\tilde{\eta} \cdot (f - \rho \ddot{u}) \mathrm{d}\Omega + \int_{\Gamma_t} H_S(x)\tilde{\eta} \cdot t \mathrm{d}\Gamma, \tag{4}$$

where $\mathcal{G}_c(\tilde{\eta}, t_{S_-})$ is the so-called contact integral.

3 FINITE ELEMENT EQUATIONS

The finite element equations emerging from variational equations (3) and (4) take the form

$$\bar{M}\mathcal{A} + F_{\mathrm{INT}}(d, d') = F_{\mathrm{EXT}}, \tag{5}$$

and

$$\tilde{M}\mathcal{A} + \mathcal{F}_{\mathrm{INT}}(d, d') + \mathcal{G}_{\mathrm{INT}}(d') = \mathcal{F}_{\mathrm{EXT}}, \tag{6}$$

where $\bar{M}$ and $\tilde{M}$ are mass matrices consistent with weighting functions $\bar{\eta}$ and $\tilde{\eta}$, $\mathcal{A}$ is the acceleration defined by $\mathcal{A} = \{\ddot{d}, \ddot{d}'\}^{\mathsf{T}}$, and d and d' are unknown nodal displacement degrees of freedom corresponding to $\bar{u}$ and $\tilde{u}$, respectively. In the penalty method, the frictional contact contribution $\mathcal{G}_{\mathrm{INT}}$ depends only on the enriched nodal degrees of freedom d'. In an explicit finite element calculation, it is desirable to obtain a lumped mass matrix $\mathcal{M}_L$. Following [5], we assume that the enriched diagonal mass terms are identical, and the conservation of kinetic energy holds. We thus obtain an enriched diagonal mass for the arbitrary enrichment function ψ:

$$m_L = \frac{1}{\sum_{i=1}^{NEN} \psi^2(x_i)} \int_{\Omega_e} \rho \psi^2 d\Omega_e. \tag{7}$$

For sign function enrichments when $\psi^2 = H_S^2(x) = 1$, the lumped mass is simply the averaged element mass over each node

$$m_L = \frac{1}{NEN} \int_{\Omega_e} \rho d\Omega_e, \tag{8}$$

where NEN is the number of nodes per element. With the above formulation, both the regular and enriched degrees of freedom can be lumped by the same classical row-sum technique when omitting the coupled terms in the consistent mass matrix in equations (5) and (6).

4 NUMERICAL EXAMPLE

In this example, we revisit Andrews' model [6] describing a rupture propagating with bulk (off-fault) plasticity. We assume a Mohr–Coulomb yield criterion for the bulk material, which is further assumed to be uniform and isotropic. We take the mass density $\rho = 2700$ kg/m^3, compressive wave velocity $V_p = 5.196$ km/s, and shear wave velocity $v_s = 3.0$ km/s. We assume the initial stress state to be uniform with $\sigma_{11} = \sigma_{22} = -50$ MPa and $\sigma_{12} = 10$ MPa. We assume that the coefficient of kinetic friction is equal to zero, and the coefficient of static friction is 0.5. For the Mohr–Coulomb model, we take cohesion $c = 0.0$, friction angle $\phi = 36.87°$, and dilation angle $\psi = 0$. In Andrews' model, a time-weakening law is used, and the characteristic time interval $T_c = 0.0035$ s. A small viscous damping coefficient $\gamma = 0.000014$ is used for this example. We define the slip zero tolerance to be equal to tol $= 4.5 \times 10^{-7}$. Penalty parameters for the penalty method and extended finite elements are $\epsilon_N = \epsilon_T = 5.0 \times 10^7$. In the computation we use a domain of 0.5 km $\times$ 8.0 km, the nucleation point is (4.0,0.25), and nucleation length $L_c = 60$ m. The element size is $h = 0.002$ km. Rupture velocity at the end of simulation is about 2.57 km/s. The contour of the plastic strain at time $t = 0.785$ s is shown in Figure 1. Slip velocity, time histories at

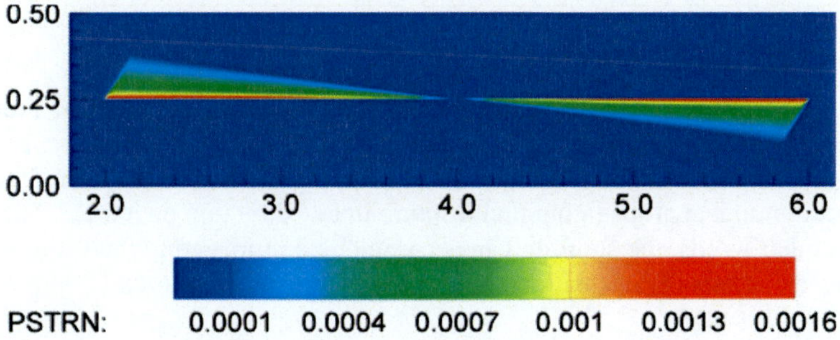

Fig. 1. Contour of equivalent plastic strain at time $t = 0.785$ s. The maximum value of plastic strain is 0.0016 near the fault surfaces.

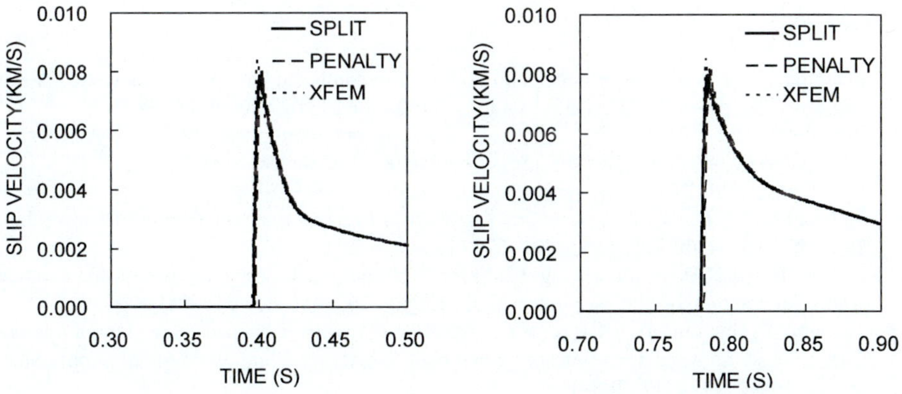

Fig. 2. Slip velocity time history at $x = 5.0$ km and $x = 6.0$ km.

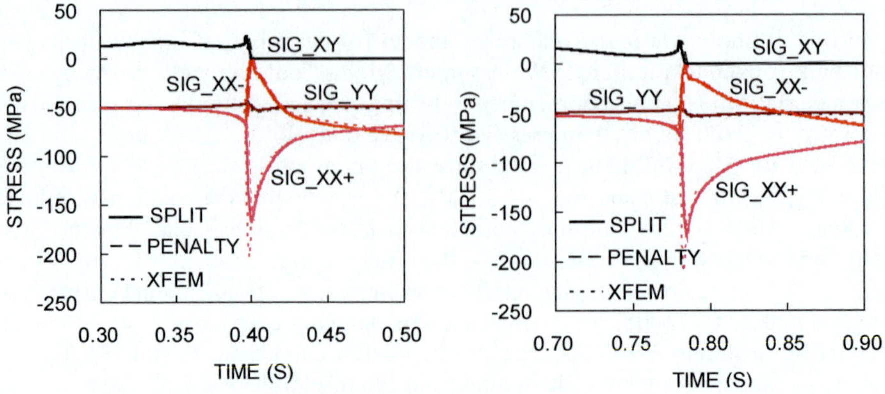

Fig. 3. Time histories of stress components at $x = 5.0$ km and $x = 6.0$ km.

$x = 5.0$ km and $x = 6.0$ km are shown in Figure 2. Stress components time histories at $x = 5.0$ km and $x = 6.0$ km are shown in Figure 3.

5 CLOSURE

The results obtained from the extended finite element simulation of fault rupture dynamics compare well with those obtained from the standard penalty and split-node techniques. Of course, the advantage of the extended finite element formulation is that it can accommodate propagating fault rupture in any direction on a fixed finite element grid. In other words, the solution can accommodate propagating fault tips without remeshing, as well as the nucleation and/or propagation of new faults.

ACKNOWLEDGMENT

This work is supported by the Office of Basic Energy Sciences, U.S. Department of Energy, under Grant No. DE-FG02-03ER15454 to Stanford University.

REFERENCES

[1] Duan, B.: Effects of low-velocity fault zones on dynamic ruptures with nonelastic off-fault response. Geophys. Res. Lett. 35, L04307 (2008), doi:10.1029/2008GL033171

[2] Templeton, E.L., Rice, J.R.: Off-fault plasticity and earthquake rupture dynamics: 1. Dry materials or neglect of fluid pressure changes. J. Geophys. Res. 113, B09306 (2008), doi:10.1029/2007JB005529

[3] Belytschko, T., Black, T.: Elastic crack growth in finite elements with minimal remeshing. Int. J. Numer. Meth. Engng. 45, 601–620 (1999)

[4] Liu, F., Borja, R.I.: A contact algorithm for frictional crack propagation with the extended finite element method. Int. J. Numer. Meth. Engng. 76, 1489–1512 (2008)

[5] Elguedj, T., Gravouil, A., Maigre, H.: An explicit dynamics extended finite element method. Part I: Mass lumping for arbitrary enrichment functions. Comput. Methods Appl. Mech. Engrg. 198, 2297–2317 (2009)

[6] Andrews, D.J.: Rupture dynamics with energy loss outside the slip zone. J. Geophys. Res. 110, B01307 (2005), doi:10.1029/2004JB003191

GEOMECHANICS APPLIED TO THE WELL DESIGN THROUGH SALT LAYERS IN BRAZIL: A HISTORY OF SUCCESS

A.M. Costa[1], E. Poiate Jr.[1], C.S. Amaral[1], A. Pereira[2], L.F. Martha[2], M. Gattass[2], and D. Roehl[2]

[1] PETROBRAS Petróleo Brasileiro S.A
Rio de Janeiro, Rio de Janeiro, Brazil
e-mail: amcta@petrobras.com.br
[2] Pontifical Catholic University of Rio de Janeiro
Rio de Janeiro, Rio de Janeiro, Brazil
e-mail: gattass@tecgraf.puc-rio.br

Summary. The lessons learned on the geomechanical salt behavior and its application in subsalt wells design are described in this article. In addition, the developed methodology validation, through comparison between computing modeling results with measurements carried out in experimental panels, in the potash mine, and with measurements obtained in an experimental well drilled for the purpose of calibrating and optimizing directional drilling in salt layers are presented. These parameters and methodology have been used for supporting the design of the wells drilled in the Pre-Salt giant oil fields in Brazil with very successful results.

Keywords: drilling, geomechanics, salt, well.

1 INTRODUCTION

A major challenge in drilling in the Pre-salt area, in Brazil, arises from the special structural salt behavior, when compared to other geomaterials, since it develops high creep strain rates under high levels of deviatoric stresses and temperatures. The salt or evaporitic rocks, formed by the sea water evaporation, have different chemical compositions. In the Pre-salt area the more important types are halite, carnallite, and tachyhydrite. The tachyhydrite, for the same state variables, deviatoric stress, and temperature, develops creep strain rates up to one hundred times higher than halite. Many operational problems, such as stuck pipe and casing collapse, have been reported when intercalation of these rocks within a thick layer is found. The challenge of designing excavations near tachyhydrite began with the development of a potash underground mine to extract sylvinite ore in Northeast Brazil. The research that began in the 1970s, to enable the mining of this ore overlying tachyhydrite, triggered one of the largest R&D projects in rock mechanics, including computing modeling, laboratory, and field tests. For the design of the pre-salt wells, this previous experience was used, and additional triaxial creep tests were performed using a new rock mechanics laboratory. Field tests and computer-modeling improvement were used to overcome the challenge of the Pre-salt drilling.

2 CONSTITUTIVE EQUATION FOR SALT BEHAVIOR

Owing to its crystalline structure, salt rocks exhibit time-dependent behavior when subjected to shear stress. The creep strain rate is influenced by the formation

R.I. Borja (Ed.): Multiscale and Multiphysics Processes in Geomechanics, SSGG, pp. 165–168.
springerlink.com

temperature, mineralogical composition, water content, presence of impurities, and the extent to which differential stresses are applied to the salt body. Early in the 1990s, creep-constitutive laws based on deformation mechanisms have been recommended by the international technical literature to represent the intrinsic behavior of the evaporates [1-2].

The law that incorporates the deformation mechanisms for the evaporite rocks was developed by Munson [1,2]. The constitutive equation based on Munson's creep law considers the following mechanisms: Dislocation Glide, Dislocation Climb, and Undefined Mechanism. The largest contribution of any of these mechanisms depends on the temperature conditions and differential stress to which the salt is submitted.

In this article, the salt rock behavior is analyzed according to the elasto/visco-elastic behavior, adopting the Double Mechanism creep law, as shown in equation 1:

$$\varepsilon = \varepsilon_0 . \left(\frac{\sigma_{ef}}{\sigma_0} \right)^n . e^{\left(\frac{Q}{RT_0} - \frac{Q}{RT} \right)} \tag{1}$$

where ε is the strain rate due to creep at the steady-state condition; ε_0 is the reference strain rate due to creep (in steady state); σ_{ef} is the creep-effective stress; σ_0 is the reference-effective stress; Q is the activation energy (kcal/mol), Q=12kcal/mol [4]; R is the Universal gas constant (kcal/mol.K), R=1.9858E-03; T_0 is the reference temperature (K); and T is the rock temperature (K).

3 VALIDATION OF THE CREEP PARAMETERS

As part of the rock mechanics studies used in enabling the mining of the lower sylvinite layer in the potash mine, an experimental panel [4] was designed and excavated in the lower sylvinite layer, overlying a layer of tachyhydrite 15m thick. An experimental room was excavated in this panel with length of 95m, divided in three sections, isolated from the effects of nearby excavations, with intensive use of field instrumentation, for back-analysis, allowing the calibration of the creep parameters. In each section, a slab protection was left with three different thicknesses (3m, 2m, and 1m). The strategy is to evaluate the influence of the slab protection thickness of sylvinite in inhibiting the floor heave due to the creep of tachyhydrite.

Among the various instruments installed in the room, this article shows the comparison between the vertical closure measurements and with those obtained by the numerical simulation. The geology description in the area of the experimental panel and the finite element model used in the simulations are shown in Figure 1. The SIGMA [4] system is used for pre- and post- processing of the finite element model. The numerical simulations have been done through application of the finite element code ANVEC [4]. Figure 2 shows the comparison among the closures measured in the experimental gallery in different locations along its axis for a slab protection of syvinite of ~2.50m. The plot shows the closure predicted by numerical simulation with and without the initial deformation after excavation, which normally is lost in the field. In these numerical models, the creep parameters obtained in the laboratory creep tests are used.

The results obtained by numerical simulation shows excellent fit to the closure measured in the experimental room C1D1 up to 16000 h. After this time, the refrigeration process starts in the mine and reduces the rock temperature and the creep measurements. This process was introduced in the numerical simulation.

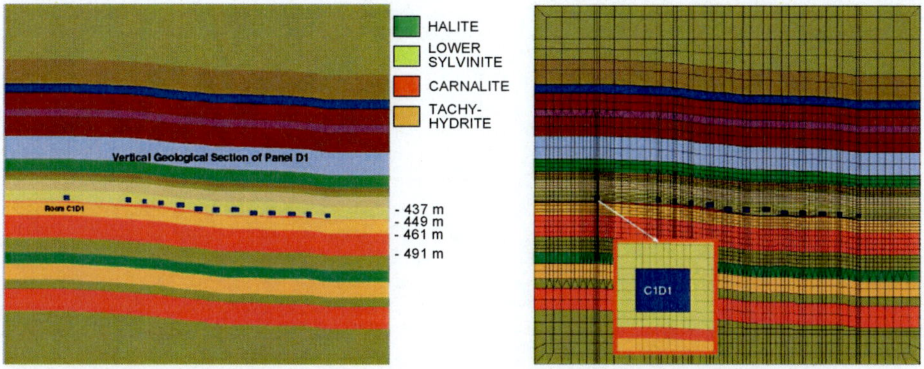

Fig. 1. Geology description in the experimental panel and the plane strain model used in the simulations.

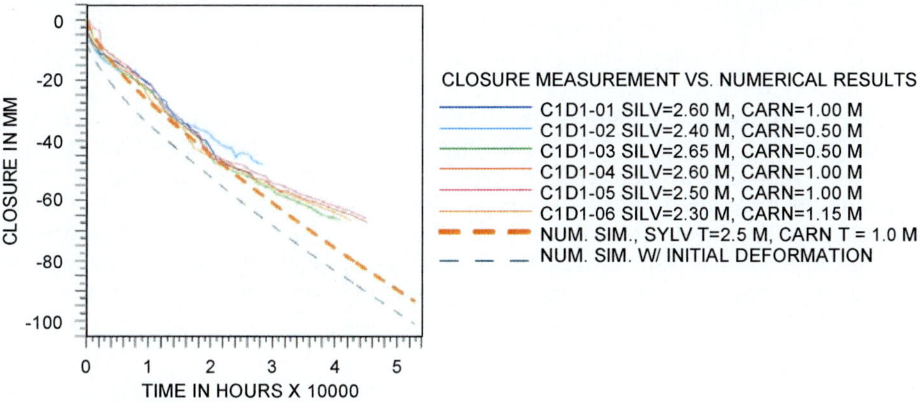

Fig. 2. Comparison between closure measurements and numerical simulation, sylvinite slap protection 2.5m thick.

4 FINITE ELEMENT MODEL OF WELL CLOSURE

With the creep parameters validation, the constitutive equation in the numerical simulation of the creep salt behavior to predict the evolution of well closure with time during drilling of thick layers of salt, was applied for various mud weights. These results were used to define several technically feasible alternatives for the drilling strategy through the salts intervals. The prospect expected 2000m of different salt rocks, to be drilled at the interval of 2600–4600m, (WD = 1600m). This case study considered a tachyhydrite layer of 5m, just 100m from the salt base. The temperature at the top of the salt interval is 40°C and at the base 60°C. The well was drilled with a diameter of 17 ½," and a 14" casing was used.

The axisymmetric model, according to the longitudinal axis of the well comprises 2000 m of salt rocks and 200 m of thick hard rock, above and below the anhydrite layer to represent the boundary condition. 86418 quadratic isoparametric elements (with 8 nodes) and 264063 nodal points are employed in the finite element model. Figure 3 shows the finite element mesh used and the numerical results, for different mud weights.

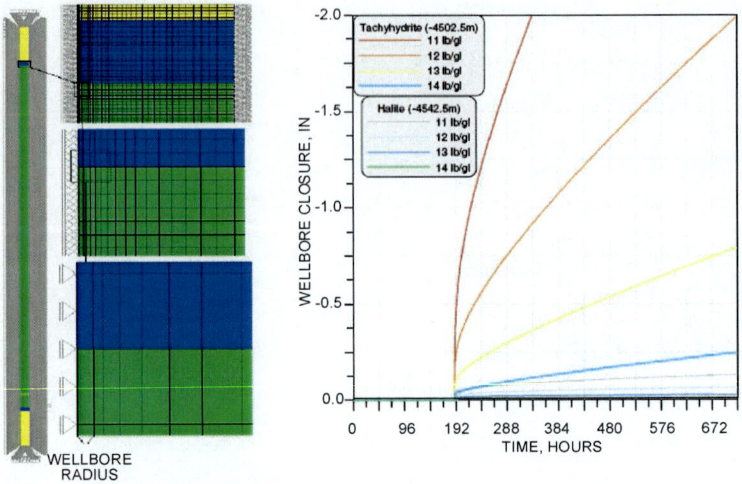

Fig. 3. Finite element mesh and numeric results.

Figure 3 shows the evolution of the 17 ½" well closure with time, when it is adopted from 11 lb/gl to 14 lb/gl mud weight. As is expected, the tachyhydrite layer has a very high closure rate, and the halite layer, a low closure rate. Considering a minimum thickness of 0.75" for the cementation of the 14" casing, for a well 17 ½" in diameter, the acceptable closure will be 2". The curves of well closures with time of each depth begin when the bit reaches the respective depth. In this case, in order to have enough time to complete the drilling operation and set the casing, use of a drilling fluid over 12lb/gl is recommended.

5 CONCLUSIONS

This study presents a methodology developed by PETROBRAS for the design of the subsalt reservoir wells for the exploration and production phase. The numeric simulations have been done through the application of an in-house-developed computer code based on the finite element method. This methodology has been used to support the design of the wells drilled in the Pre-Salt giant oil fields in Brazil with very successful results.

REFERENCES

[1] Munson, D.E., Fossum, A.F., Senseny, P.E.: Approach to first principles model prediction of measured wipp (Waste Isolation Pilot Plant) in-situ room closure in salt. Tunneling and Underground Space Technology 5, 135 (1990)

[2] Munson, D.E., Devries, K.L.: Development and validation of a predictive technology for creep closure of underground rooms in salt. In: Seventh International Congress on Rock Mechanics, Deutschland, vol. 7 (1991)

[3] Frayne, M.A., Mraz, D.Z.: Calibration of a numerical model for different potash Ores. In: Proceedings of the Seventh International Congress on Rock Mechanics, Deutschland, vol. 7 (1991)

[4] Costa, A.M., Poiate Jr., E., Claudio, S.A., et al.: Geomechanics applied to the well design through salt layers in Brazil: A History of success. In: 44th US Rock Mechanics Symposium and 5th U.S. Canada Rock Mechanics Symposim, Salt Lake City, UT, June 27-30 (2010)

LANDSLIDE AND DEBRIS FLOW-INDUCED STATIC AND DYNAMIC LOADS ON PROTECTIVE STRUCTURES

Carter M. Mast*, Peter Mackenzie-Helnwein, Pedro Arduino, and Gregory R. Miller

Department of Civil and Environmental Engineering
201 More Hall, Box 352700
University of Washington, Seattle, WA 98195-2700
* e-mail: cmast@u.washington.edu, web page: http://www.ce.washington.edu

Summary. This study presents two suitable extensions to the Material Point Method (MPM) for improved modeling capabilities in the context of landslide and debris flows. The first provides a multi-field formulation strategy to support solid/fluid phases and multi-scale interactions, while the second provides the ability to represent a general bounding surface within the MPM framework.

Keywords: boundary conditions, dual-grid, meshfree methods, MPM, multi-mesh.

1 INTRODUCTION

Landslides and debris flows cause significant damage and loss of life around the world each year. In the US alone, the annual economic costs of landslides can be estimated conservatively to be between \$1 and \$2 billion, with an associated 25–50 yearly casualties [1]. To help protect people, infrastructure, and lifelines against such effects, it is critical to have engineered structures that are capable of resisting the loads that are induced during these events. However, determining such loads can be challenging—landslides and debris flows are highly dynamic events and inherently complex in nature. The true extent of this complexity can be better understood by considering some of the key challenges that arise in attempting to model a landslide or debris flow. A general implementation must include the ability to capture the transitions between solid-like and fluid-like states, phase interaction modeling, appropriate volume constraints, boundary condition generalization, and constitutive modeling, all within the context of motions with complexities on many scales. The purpose of this study is to develop a robust numerical modeling framework capable of accommodating these key phenomena with the primary initial goal of predicting loads on protective structures. The basic approach is based on adapting and extending the Material Point Method (MPM), which is well suited to modeling large deformation, flow-like phenomena. This article includes a brief description of the standard MPM algorithm, and then focuses on the two extensions to the method implemented to date: (i) a multi-field formulation strategy to support solid/fluid phases and interactions; and (ii) boundary condition generalization.

R.I. Borja (Ed.): Multiscale and Multiphysics Processes in Geomechanics, SSGG, pp. 169–172.
springerlink.com

Fig. 1. Simulated transition from static behavior to slide initialization and subsequent debris flow on a rough plane slope.

2 APPROACH

The MPM is a numeric technique that has its roots in fluid mechanics and was extended for use in solid mechanics by Sulsky et al. [2]. In this method, the governing equations are solved at nodal points on an Eulerian grid. Each generic body is represented as a collection of free moving *material points*, frequently referred to as *particles*. State variables are tracked at the particle level in a Lagrangian manner. The hybrid Eulerian–Lagrangian description allows large deformation solid-like behavior and fluid-like flow to be accurately represented in a single framework—a critical feature needed for proper analysis of landslides and debris flows. While the standard algorithm contains several key features that make it an ideal tool for this type of analysis, the method is not without its shortcomings or pitfalls. In particular, without modification, the standard implementation lacks a consistent multi-field formulation for use in a landslide or debris flow context, and the method does not have the ability to represent a general bounding surface. These obstacles can be overcome by providing suitable extensions to the standard algorithm. Figure 1 shows an example landslide simulation generated using one such enhanced MPM framework.

2.1 Multi-scale Modeling and Multi-field Formulation

One of the key challenges in multi-phase flow analysis is the development of an accurate interaction model describing the behavior between two or more phases. This challenge is composed primarily of length scale considerations and how each phase is treated (i.e., as a distinct phase, volume averaged, etc.) within the flow. For landslides and debris flows, the length scale of the event itself is several orders of magnitude larger than the characteristic length scale of the individual constituents. This sets the stage for a multi-scale analysis effectively linking the micro-structure phase interaction model with the macroscale response. This concept is illustrated in Figure 2 for two distinct phases (i.e., f and s for fluid and solid phases, respectively). This approach assumes that each phase is isolated at both the micro-structure and the macro-structure level and can thus be assigned to its own motion. Moving up in the figure illustrates localization, whereas moving down illustrates homogenization.

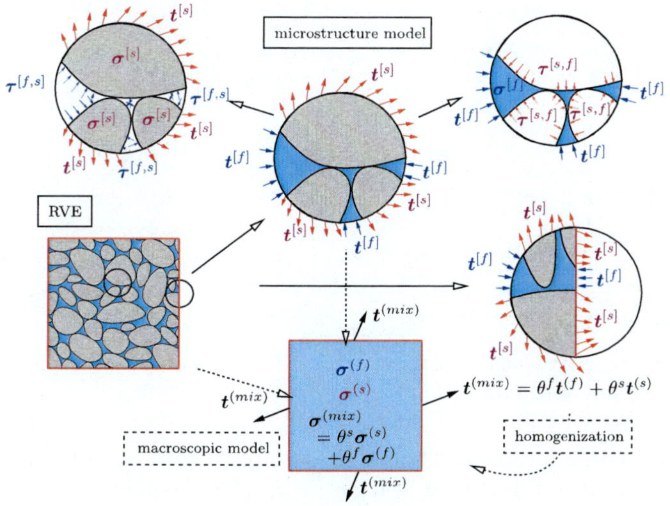

Fig. 2. Illustration of the multi-scale model for a fluid–solid two-phase mixture using distinct phases.

Quantities $(\bullet)$ associated with a phase α at the micro-structure level are identified using square brackets $((\bullet)^{[\alpha]})$, while those associated with the macro-scale representation are identified using parentheses $((\bullet)^{(\alpha)})$.

The link between the microscopic stresses $\boldsymbol{\sigma}^{[\alpha]}$ (or the respective interface traction $\boldsymbol{\tau}^{[\beta,\alpha]}$) and the respective macroscopic quantities $\boldsymbol{\sigma}^{(\alpha)}$ and $\boldsymbol{b}^{(\beta,\alpha)}$ of phase α are obtained through volume averaging over a representative volume element (RVE). The geometry of an RVE is ideally chosen such that any averaged quantity is stationary with respect to changes in size and shape of the RVE. In fact, every single particle is viewed as an RVE carrying the averaged quantities for the volume it represents. In this sense (i.e., using both the particle and the cell as RVEs), the discussed concept provides a suitable framework for multi-scale modeling.

Once the macroscopic quantities $\boldsymbol{\sigma}^{(\alpha)}$ and $\boldsymbol{b}^{(\beta,\alpha)}$ are obtained, different techniques are available for modeling the drag interaction between two (or more) phases. In Mackenzie-Helnwein et al. [3], each phase is treated with its own motion, and different families of phase interaction strategies are explored within the context of the MPM. Their study considers both nodal- and particle-based approaches, and in particular, a new smoothed volume fraction is presented. This multi-field formulation allows for a comparably simple simulation of mixing, interaction, and separation of phases.

2.2 Dual-Grid Extension

A critical component needed for modeling landslides and debris flows is the ability to represent a general bounding surface. A general representation serves two purposes: (i) it allows naturally occurring topological features to be incorporated into the model, and (ii) it allows the geometry of protective structures to be represented. The latter is required for successful recovery of the loads due to flow–structure interaction.

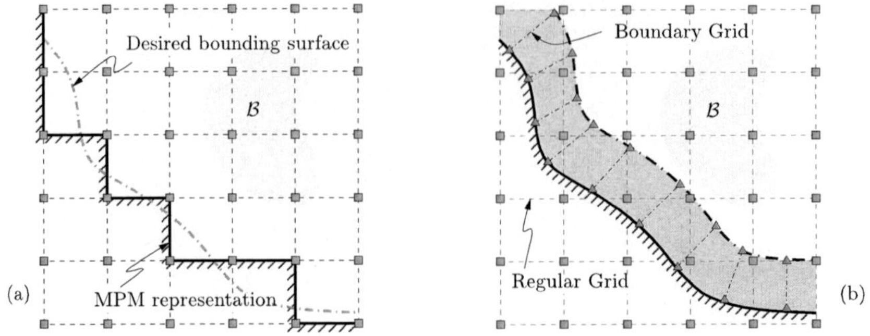

Fig. 3. (a) Boundary representation in the standard MPM and (b) Dual-grid approach for enforcing arbitrary boundary conditions.

The standard algorithm models a bounding surface by restricting nodal values in the computational domain. For regular grid geometries (as is commonplace in MPM analyses), this leads to significant limitations on the shape of a boundary, as shown in Figure 3a for a general body, $\mathcal{B}$. It is apparent that general surface geometry is reduced to a combination of horizontal and vertical planes when a regular grid is used. While for some applications, this restriction may match the desired surface, in general, this offers an unrealistic representation. The dual-grid extension allows general boundary features to be incorporated by introducing a separate, independent *boundary grid* that conforms to the desired geometry as shown in Figure 3b. The regular and boundary grid dynamics are linked via a modified set of weak form equations. In Mast et al. [4], two distinct methods are presented within the dual-grid framework. The first enforces a strong-form boundary condition with weak grid coupling, while the second satisfies a weak form boundary condition with directly coupled grid dynamics. The effectiveness of each proposed method is demonstrated using an example problem.

REFERENCES

[1] Highland, L.M.: Estimating landslide losses—preliminary results of a seven-state pilot project. Technical Report 2006:1032, USGS (2006), `http://pubs.usgs.gov/of/2006/1032/pdf/OFR-2006-1032.pdf`

[2] Sulsky, D., Zhou, S., Schreyer, H.L.: Application of a particle-in-cell method to solid mechanics. Computer Physics Communications 87(1-2), 236–252 (1995)

[3] Mackenzie-Helnwein, P., Arduino, P., Shin, W., Moore, J.A., Miller, G.R.: Modeling strategies for multiphase drag interactions using the material point method. International Journal for Numerical Methods in Engineering, 83(3), 295–322 (2010)

[4] Mast, C.M., Mackenzie-Helnwein, P., Arduino, P., Miller, G.R.: Representing arbitrary bounding surfaces in the material point method using a dual-grid approach. International Journal for Numerical Methods in Engineering (in review, 2011)

IMPLEMENTATION OF MULTISCALE MODELS IN A PROBABILISTIC FRAMEWORK FOR PERFORMANCE-BASED DURABILITY ENGINEERING

Madeleine M. Flint* and Sarah L. Billington

Department of Civil and Environmental Engineering
Stanford University
John A. Blume Earthquake Engineering Center
MC: 4020, Building 540
Stanford CA 94305-4020
*e-mail: flint@stanford.edu
e-mail: billington@stanford.edu, web page: http://www.stanford.edu/~billingt/

Summary. Corrosion of reinforcing bars in concrete presents a global problem for safety, serviceability, and economic and environmental sustainability. Owing to the interrelation between transport of corrosives and cracking, predicting the onset and consequences of corrosion of steel reinforcement is challenging. Multiscale models of hydration, transport of water and chlorides, cracking, and material degradation can be used to improve predictions of structural capacity and remaining service life. Implementation of such models within a probabilistic framework that allows incorporation of uncertainty as well as analysis of corrosion-related outcomes of interest to owners is being studied. The performance-based durability engineering (PBDE) framework presented here combines flexibility in modeling with industry-oriented decision-making information.

Keywords: corrosion, durability, multiscale modeling, reliability.

1 INTRODUCTION

Existing frameworks assess durability from the perspective of reliability or design. Reliability assessment of corroding concrete structures and components has been well researched [1-3, among others]. These assessments make use of simplified diffusion and corrosion models and concentrate on incorporating material and modeling uncertainty in predictions of service life and safety. Design frameworks, such as ACI Life-365, DuraCrete, and FIB Model Code Bulletin 34, use a mix of empirical and theoretical models to assess concrete mix designs and structural geometry for resistance to a variety of degradation mechanisms [4-6]. Owners require information for both design and assessment, which suggests that a more general framework will be useful.

2 PROPOSED PBDE FRAMEWORK

The proposed PBDE framework is modeled on a performance-based earthquake engineering framework [7]. The PBDE framework moves through four stages of

R.I. Borja (Ed.): Multiscale and Multiphysics Processes in Geomechanics, SSGG, pp. 173–176.
springerlink.com

analysis, from assessment of the site and structure to likely economic losses and other impacts associated with corrosion. The four stages, shown in Figure 1, are (1) exposure analysis to determine the exposure conditions (EC) for the structure; (2) deterioration analysis that predicts the onset of corrosion, rust penetration and cracking as damage measures (DM); (3) maintenance and repair analysis that uses the damage type and degree to predict the probability of certain repair actions (RA); and (4) impact analysis that links repair actions to their costs, downtime, and social and environmental impacts, referred to as decision variables (DV). Multiscale transport and damage models are incorporated in the deterioration analysis.

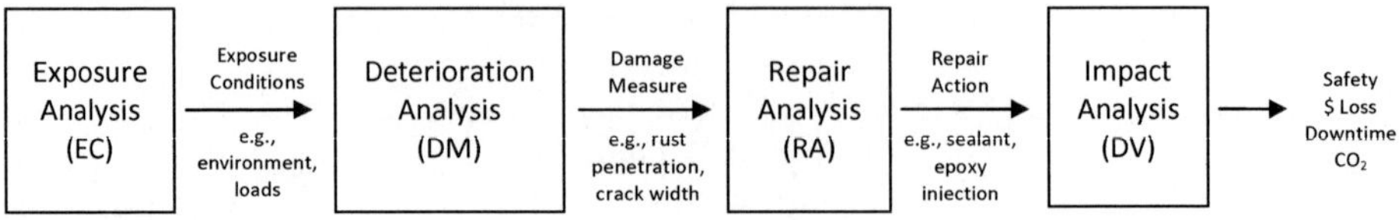

Fig. 1. PBDE framework

By integrating over the range of exposures and uncertainty in the modeling stages it is possible to predict rates of exceeding a value of a decision variable (v_{DV}), using Equation 1.

$$v_{DV}(DV) = \iiint G(DV|RA)\, dG(RA|DM)\, dG(DM|EC)\, dG_{EC} \quad (1)$$

As in performance-based earthquake engineering, the stages of analysis are separated so that exceedence curves can be obtained at intermediate points. Depending on the desired outputs, this integration can be performed continuously in time, or cumulatively for a design life, as follows:

(1) Exposure analysis determines the suite of exposure condition variables (EC). These ECs include surface chloride concentration, temperature, humidity, and loads during the analysis period considered. Depending on the level of detail required, these variables may be modeled as stationary, periodic, or non-stationary stochastic processes. ASTM guides and maps showing atmospheric corrosive potential can be adjusted to reflect corrosion potentials for reinforced concrete, similar to the USGS seismic hazard maps [8]. More unique exposure conditions, such as structures in marine splash zones that are subjected to de-icing salts in winter, will require site-specific analysis.

(2) Structural and deterioration analysis links the ECs to related damage measures in the structure. As water and chloride transport and cracking occur in the structure, the capacity is updated to reflect degradation. This allows calculation of structural reliability and safety, as well as serviceability checks such as crack width limits. It is in this stage of the analysis that multiscale and multiphysics models are used. The potential implementation of such models is discussed further in Section 3. Damage measures include depth of chloride ingress, volume or mass of corroded steel reinforcement, pit depth, and crack width.

(3) Repair analysis ties DMs obtained from the deterioration analysis to the expected repair actions. These repair actions may be customized to reflect the preferred rehabilitative strategies of the structure's owners or operators. Repair actions might include placement of a sealant, cathodic protection, replacement of

concrete cover and rebar, or external FRP jacketing. New fragility curves relating a DM to the probability of implementing repair actions need to be developed. If a repair is expected to alter transport or structural properties, then the structural and deterioration model can be updated to reflect these changes, allowing a more refined estimation of post-rehabilitation safety and service life.

(4) Impact analysis utilizes the repair actions and their associated probabilities to calculate dollar losses, downtime, environmental impact, and system safety and reliability. A database of repair actions and their distributions of required materials, costs, and duration can be linked to life cycle assessment tools to predict environmental impacts, such as the releases greenhouse gases. Some loss data are already available through earthquake-engineering toolkits, and commercial software for life cycle assessment can be used for environmental impact assessment [9]. The output variables obtained are expected to enhance decision making with regard to optimization of new design, rehabilitative strategy, or maintenance scheduling.

Furthermore, the expected values and distributions of these impacts will be used in assessing the effects of using different models within the PBDE framework. For example, if a simple model gives impact results similar to those of a more complex, computationally expensive model, the former could be adopted for use in industry. Analysis of the main contributors to the impacts will reveal where uncertainty in model inputs has a large effect on the predicted outcomes, providing motivation for future research. Variables that do not greatly affect decision variables may be modeled as deterministic in a simplified approach.

3 INTEGRATION OF MULTISCALE AND MULTIPHYSICS MODELS

Multiscale and multiphysics models are incorporated into the structural and deterioration analysis of stage 2. Models are currently available for three main phases of the deterioration analysis: cement hydration, water and chloride transport, and finite element structural analyses incorporating cracking.

Concrete hydration models are available in DuCOM, Hymostruc, and CEMHYD3D [10-12]. The models use concrete mix designs to create stochastic representations of concrete rheology to calculate macroscopic material properties as functions of concrete maturity. Transport properties (e.g., diffusivity) and mechanical properties (e.g., stiffness, compressive strength) will be linked to finite element analyses of the degrading structure.

The three hydration models also have capabilities of modeling water and chloride transport in intact concrete. Transport of chlorides occurs due to absorption, permeation, or diffusion in saturated media. While these phenomena are sometimes grouped into "apparent" Fickian diffusion for the purposes of predicting corrosion onset, recent research offers improved models for transport in damaged porous media. Recent research by Abreu, et al. [13] and Kemper, et al. [14], among others, model the transport behavior in cracked materials, which is essential for accurately capturing the macrocell corrosion actually seen in deteriorating structures.

As performing a microscale transport analysis of the entire concrete structure is prohibitively and computationally intensive, a multiscale approach to macroscopic behavior is required. Finite element structural models incorporating hydration model parameters will be used for predicting cracking under service loads. While bulk transport properties can be assumed in uncracked regions, the cracked regions will be modeled at a mesoscale using the multiphysics transport models. Mesoscale damage due to the volumetric expansion of corrosive products will then be used to update the macroscale model by reducing bond coefficients and applying internal displacement or pressure in the corroding regions.

4 CONCLUSIONS

Multiscale and multiphysics models are being incorporated into a probabilistic framework for performance-based durability engineering. The use of these models is expected to result in more robust predictions of structural deterioration, safety, loss, and environmental impacts.

REFERENCES

[1] Enright, M., Frangopol, D.: Probabilistic analysis of resistance degradation of reinforced concrete bridge beams under corrosion. Engineering Structures 20(11), 960–971 (1998)

[2] Thoft-Christensen, P.: Assessment of the reliability profiles for concrete bridges. Engineering Structures 20(11), 1004–1009 (1998)

[3] Li, C.Q., Melchers, R.E.: Time-dependent reliability analysis of corrosion-induced concrete cracking. ACI Structural Journal 102(4), 543–549 (2005)

[4] Thomas, M., Bentz, E.: Life-365: Computer program for predicting the service life and life-cycle costs of reinforced concrete exposed to chlorides. ACI (2000)

[5] Siemes, T., Edvardsen, C.: Duracrete: service life design for concrete structures. In: Proceedings, vol. 8, pp. 1343–1356. DBMC, Vancouver (1999)

[6] Fédération Internationale du Béton. Bulletin 34: Model Code for Service Life Design,fib, Lausanne (2006)

[7] Deierlein, G., Krawinkler, H., Cornell, C.: A framework for performance-based earthquake engineering. In: Proceedings of 2003 Pacific Conference on Earthquake Engineering, New Zealand (2003)

[8] ASTM G101, Standard guide for estimating the atmospheric corrosion resistance of low-alloy steels, ASTM (2010)

[9] Hamburger, R.: The ATC 58 Project: Development of next generation performance based earthquake engineering design criteria for buildings. In: Proceedings, Structures Congress 2006, St Louis, ASCE (2006)

[10] Maekawa, K., Ishida, T.: Service-life evaluation of reinforced concrete under coupled forces and environmental actions, pp. 219–238. University of Tokyo, Tokyo

[11] Van Breugel, K.: Simulation of Hydration and Formation of Structure in Hardening Cement-Based Materials—HYMOSTRUC. Ph.D. Thesis, 2nd edn. Delft University of Technology (1997)

[12] Bentz, D.: CEMHYD3D: A three-dimensional cement hydration and microstructure development modelling package, NIST (2000)

[13] Abreu, M., Carmeliet, J., Lemos, J.V.: Prediction of the permeability of damaged concrete using a combined lattice beam-crack network approach. In: Proceedings, EURO-C 2010, Austria, pp. 431–440 (2010)

[14] Kemper, M., Timothy, J.J., Kruschwitz, J., Meschke, G.: Modeling of Chloride and CO2 transport in intact and cracked concrete in the context of corrosion predictions of RC structures. In: Proceedings EURO-C 2010, Austria, pp. 503–512 (2010)

MULTIPHYSICAL SIMULATIONS OF THE FREEZING EFFECTS ON SOILS

Zhen Liu[1] and Xiong (Bill) Yu[2]

[1] Graduate Research Assistant, Department of Civil Engineering,
Case Western Reserve University,
10900 Euclid Avenue, Bingham 256, Cleveland, OH 44106
e-mail: zhen.liu@case.edu, web page: http://filer.case.edu/zxl73/
[2] Assistant Professor, Department of Civil Engineering,
Case Western Reserve University,
10900 Euclid Avenue, Bingham 256, Cleveland, OH 44106
e-mail: xxy21@case.edu, web page: http://filer.case.edu/xxy21/Index.html

Summary. This article describes the development and implementation of a coupled thermo-hydro-mechanical model to study the soil–structure interactions under frost effects. The model incorporates unsaturated soil mechanics and governing principles to establish the thermo-hydraulic coupling schema. Simulations were conducted on soil–structure interactions under freezing. A case study is presented to compare the simulation results with testing data obtained from a road section of the Strategic Highway Research Program. The boundary and initial conditions were set from field instrumentation data. The predicted subsurface environmental field was compared against the actual measured data. The results of simulation match very well with the field measured data, including both the subsurface temperature and water content distribution. This article demonstrates the power of simulation model to investigate the complex interactions between soil and infrastructure under the freezing effects.

Keywords: freezing, multiphysics, thermo-hydraulic-mechanical coupling.

1 INTRODUCTION

Multiphysical processes are responsible for many phenomena observed in unsaturated porous materials. Coupling of different fields, such as thermal field or hydraulic field with stress field, can easily lead to noticeable stress or deformation without external loads. A common problem involving the coupling processes is the frost-induced ground heave. This phenomenon is caused by the coupled hydro-thermal procedures upon ground freezing–thawing. The resultant differential settlement causes a variety of damages to pavement, pipes, etc.

This study is conducted in view of two major needs to advance mechanics of freezing ground: 1) while there exist research studies on the coupled thermo-hydro-mechanical process in porous materials such as concrete or rock, less attention has been placed on soils; and 2) most of the current models are based on empirical assumptions rather than the fundamental principles. The final goal of this study is to develop and implement holistic multiphysical simulations for freezing soils.

2 GENERAL SPECIFICATIONS

The coupling of the three physical fields were achieved through variables and parameters that transfer information interactively. Owing to high nonlinearity, the equations have to be solved numerically. The multiphysical coupled model is solved with assistance of COMSOL, a commercial multiphysical simulation platform. The software provides an interactive graphic interface to construct the geometry model. The model is then custom built and solved using the nonlinear solver provided by COMSOL. The results can then be analyzed using the post-processing functions.

A simulation was conducted using the geometry and boundary conditions similar to that used by Mizoguchi (1990). A plane strain computational domain with 10-cm length, 8-cm height is used. The thermal boundary conditions are shown in Figure 1. The heat influx is governed by Newton's law of cooling:

$$\vec{\mathbf{n}} \cdot (\lambda \nabla T) = \mathrm{h_c}(T_{\mathrm{emb}} - T) \tag{1}$$

where T_{emb} is the ambient temperature, and T is the surface temperature of soil specimen.

All boundaries were hydraulically insulated to ensure mass conservation. The mass conservation equation is described as

$$\vec{\mathbf{n}} \cdot (K_{\mathrm{Lh}} \nabla h + K_{\mathrm{Lh}} + K_{\mathrm{LT}} \nabla T) = 0 \tag{2}$$

For the stress field, boundary condition is to counterbalance the initial gravity field. The gravity field was considered by imposing an initial strain.

To implement the multiphysical simulation, the differential equations were first transformed into weak forms.

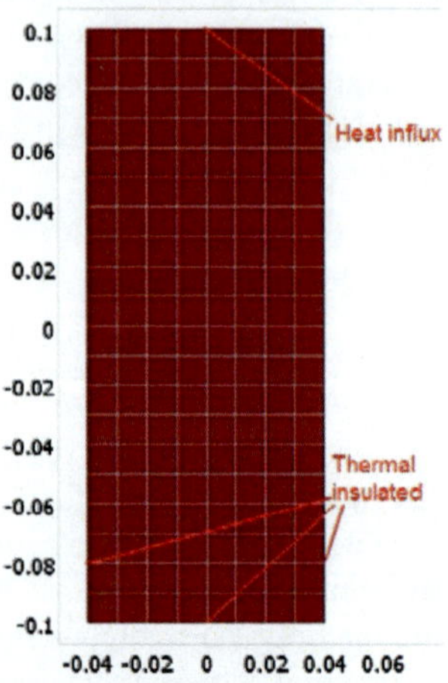

Fig. 1. FEM mesh of computational domain with thermal boundary conditions

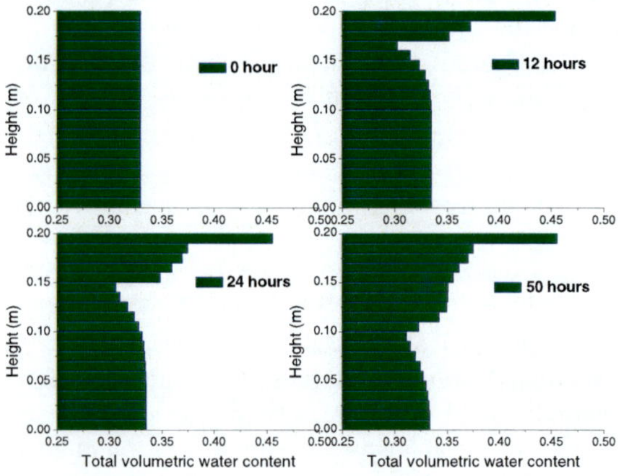

Fig. 2. Distribution of the total volumetric water content at different times

3 EXAMPLE RESULTS

Figure 2 shows the plots of the distribution of the total volumetric water content (the volumetric water content plus the volumetric ice content) at different times. The curves clearly demonstrate the trend of moisture migration toward the frost front. The depth of frost penetration can be approximately identified as the location with the lowest liquid water content.

Figure 3 shows the distribution of vertical stress in the specimen due to freezing. Both the maximum tension and compression stresses appear at the surface layers. The reason is that the moisture will migrate and accumulate near the surface. This caused volume expansion and the internal stresses.

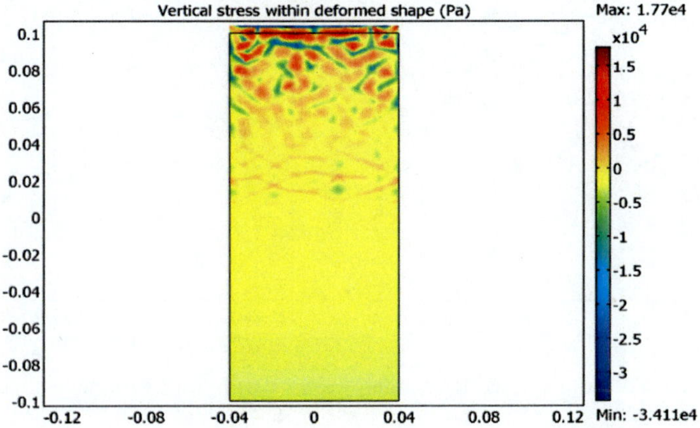

Fig. 3. Distribution of internal stress under freezing

The model was validated by simulating the distribution and variation of temperature and moisture content under pavement; both are important for pavement performance. The Ohio SHRP test pavement was selected as the testbed. The environmental fields, such as the temperature and moisture content distribution under the base and subgrade, were monitored. Besides, the air temperature and precipitation have been measured since 1996. A test section of asphalt pavement (F8) was selected from the test road. The section includes 8 inches asphalt concrete pavement and 6 inches aggregate base. Eighteen temperature gauges and 10 moisture gauges were installed within the pavement.

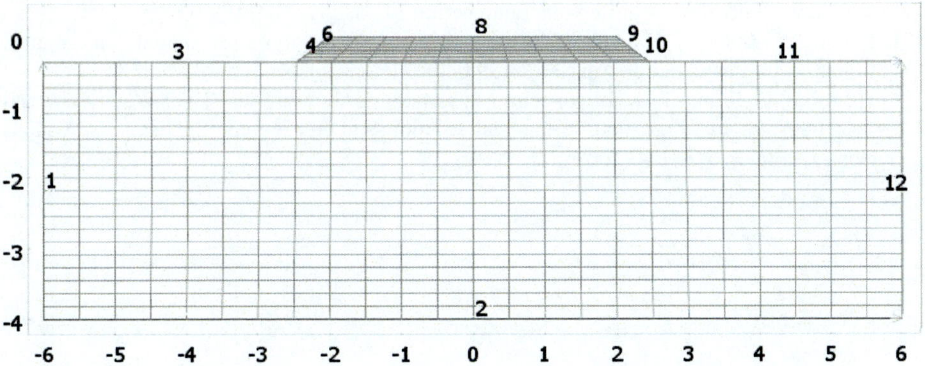

Fig. 4. Meshed computational domain and boundary

The simulation was conducted from December 10 to December 19, 2000. Figure 5a illustrates the comparison between the measured and simulated temperature variation at the depth of temperature gauges S1, S2, and S5 (The depths of these temperature gauges were 1 inch, 4 inches, and 13 inches below the surface of road, respectively). Figure 5b shows the distribution of volumetric total water content at different times.

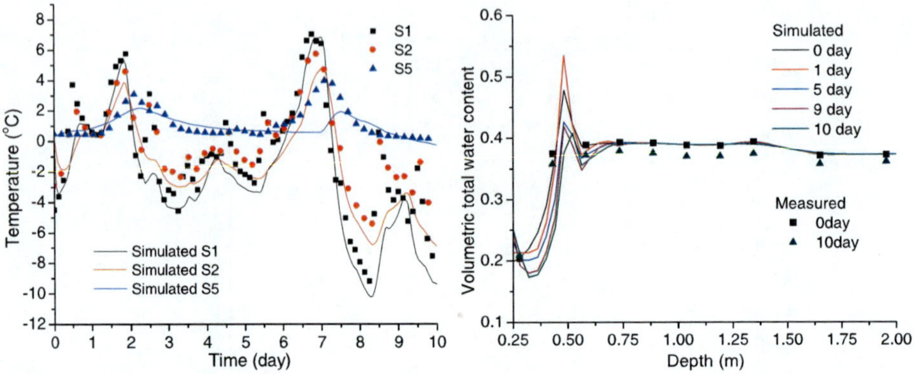

Fig. 5. a) Simulated and measured temperature varying with time; b) Simulated and measured moisture distribution

4 CONCLUSIONS

A multiphysical model for freezing soils is developed based on basic laws for heat balance, mass balance, and the elasticity theory. Relationships, such as the water intension curve and ice–water balance described by the Clapeyron equation, were introduced to provide framework system of equations. The highly non-linear equations were solved numerically in a multiphysical finite element simulation platform. Results are consistent with the experimental observations.

REFERENCES

[1] Simonsen, E., Janoo, V.C.: Prediction of temperature and moisture changes in pavement structures. Journal of Cold Regions Engineering 11(4), 291–307 (1997)

[2] Rajani, B., Zhan, C., Kuraoka, S.: Pipe-soil interaction analysis for jointed water mains. Canadian Geotechnique Journal 33(3), 393–404 (1996)

[3] Hansson, K., Simunek, J., Mizoguchi, M., Lundin, L.C., Martinus, T.G.: Water flow and heat transport in frozen soil: Numerical solution and freeze-thaw applications. Vadose Zone Journal 3, 693–704 (2004)

CENTRIFUGE MODEL TEST AND NUMERICAL ANALYSIS ON FACE STABILITY OF SHALLOW TUNNELS

Gregor Idinger, Pelin Aklik, Xiaoyu Song, and Wei Wu

Institute of Geotechnical Engineering
University of Natural Resources and Applied Life Sciences
Feistmantelstraße 4., 1180 Vienna, Austria
e-mail: gregor.idinger@boku.ac.at
web page: http://www.baunat.boku.ac.at/489.html

Summary. In this article, we report on model tests of shallow tunnel in a geotechnical centrifuge. The influence of the overburden on the required face support pressure, ground deformation, failure mechanism and surface settlement is investigated. Owing to small shear strain in pre-failure regime, high spatial resolution is needed in the instrumentation. The experimental results are then compared with the closed form solutions and a finite element analysis.

Keywords: centrifuge test, face stability, PIV, shallow tunnel.

1 INTRODUCTION

Shallow tunnels in soft ground are often driven by a shield TBM, which use pressurized materials (e.g. spoils) as face support. The estimation of the required support is a topic of current research. Available theoretical models provide a wide range of the earth pressure. According to the theoretical model, if active earth pressure is balanced by support pressure, no deformations will occur. However, higher support pressure gives rise to construction cost and time. Moreover, a high support pressure may induce blow-outs and heave of the ground. Experience has shown that a reduced support pressure can be successfully used without inducing excessive ground deformations [1]. However, the extent of the ground movements associated with lower pressure is not yet well known.

2 METHODS

In order to achieve stress similitude, the present experiments are performed on a small scale model in a geotechnical centrifuge at 50 g. The centrifuge has a diameter of 3.0 m, a load capacity of 100 kg at 100 g, and maximum gravitational acceleration of 200 g.

The measurement of ground deformation is important to understand the failure mechanism. Usually, the failure pattern is ascertained by excavating the soil at test end. However, initial ground deformation occurs at a much lower strain level (0.01–1.0 %). To quantify small displacement, Particle Image Velocimetry (PIV) was used [2]. This technique enables displacement measurement with high resolution and accuracy over the entire domain. The use of PIV is feasible, provided the surface contains sufficient texture. Natural sand fulfils this requirement [3]. PIV analysis is applied with a maximum uncertainty of 0.03mm (0.01%).

R.I. Borja (Ed.): Multiscale and Multiphysics Processes in Geomechanics, SSGG, pp. 181–184.
springerlink.com

3 EXPERIMENTAL SET-UP

A tunnel model with a circular cross section is considered. To facilitate displacement measurement, we make use of symmetry and examine only half of the tunnel (Figure 1). The plane of symmetry is along transparent perspex with thickness of 30 mm. The model box has a height of 400 mm, is 440 mm in width and 155 mm in depth. Tunnel lining is modeled by a cylindrical half-shell with a diameter of 100 mm and a thickness of 3 mm. The support is provided by a rigid piston, connected to a linear actuator. Ground collapse is induced by driving back the piston. System friction between piston and half-shell is determined by a calibration test without sand.

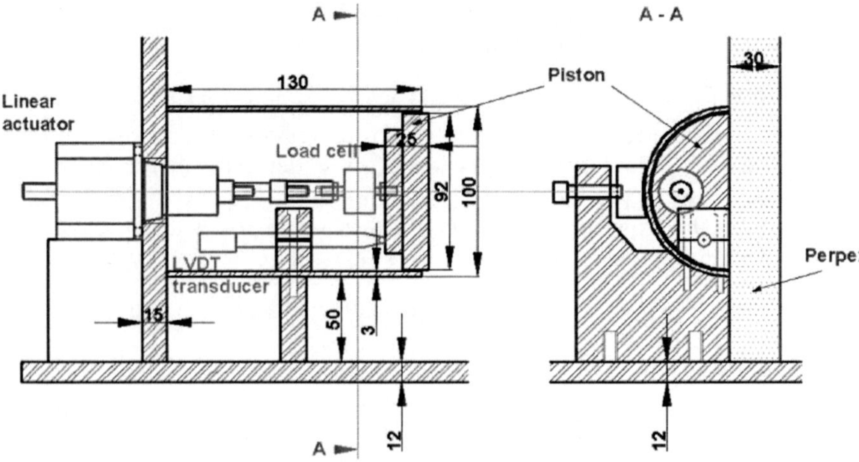

Fig. 1. Schematic diagram of the tunnel model, frontal view and cross-section

4 NUMERICAL INVESTIGATION

In addition to the experimental tests, a numerical analysis is performed with the commercial code FLAC3D. A three-dimensional analysis is necessary because ground movements are dependent on the distance to the tunnel axis.

5 RESULTS

Centrifuge model tests and numerical analysis are performed for three different tunnel cover-to-diameter ratios C/D: 1.5 (T1, T2), 1.0 (T3, T4, T5), and 0.5 (T6). Medium sand with a mean grain diameter of about 0.8 mm is used. The sand is deposited by air pluviation.

Figure 2 shows the results of the support pressure. The influence of the overburden can be observed. At the beginning of the test, the measured pressure drops remarkably for small piston displacement. For increasing displacement, the support pressure reduces gradually, to reach 5–10 kN/m². These values can be interpreted as the minimal pressure for face support.

The result of the PIV analysis for C/D=0.5 is presented in Figure 3. Figure 3a illustrates the displacement field caused by a piston displacement of about 5 mm. In the shear strain field of Figure 3b, two slip surfaces can be observed. The primary failure surface develops from the tunnel base and propagates under an angle of 60-70°

toward ground surface. Reaching the elevation of the tunnel crown, the failure surface propagates almost vertically until ground surface, parallel to a second failure surface arising at the tunnel crown.

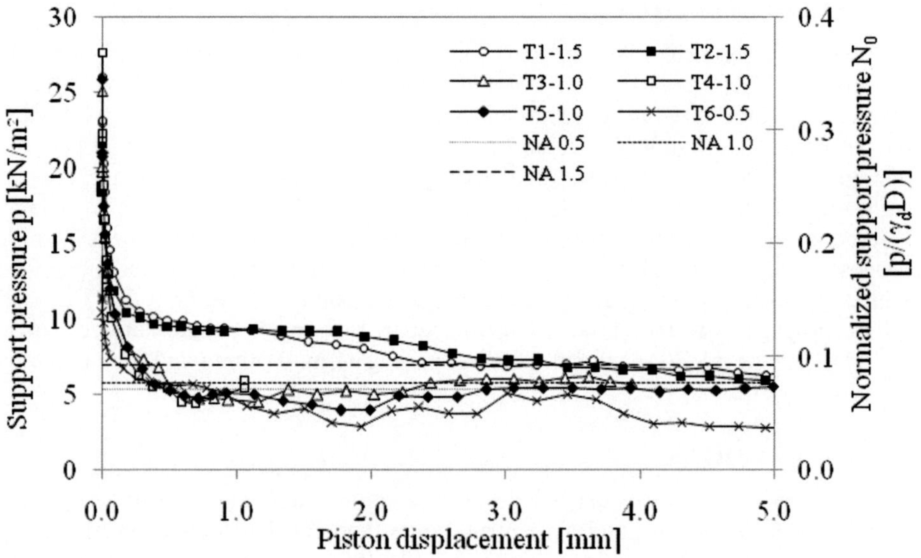

Fig. 2. Support pressure over a piston movement of 5mm; straight lines represent numeric analysis

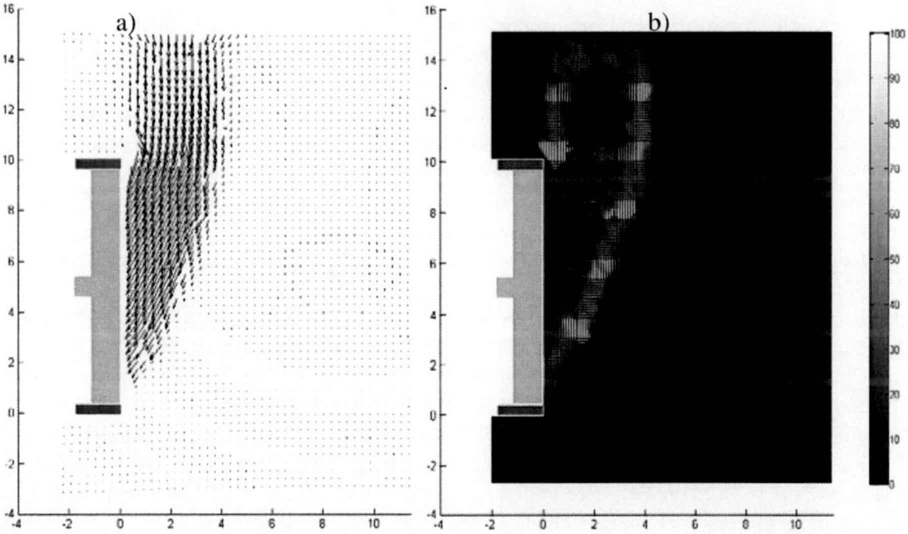

Fig. 3. Results of PIV analysis for T6; axes unit: centimeter; a) displacement vectors in a scale of 1 b) contours of shear strain; scale 0–100%

The surface settlement in the longitudinal tunnel section is obtained from the PIV data. At the final piston movement of 5 mm, maximum settlement is about 1 mm for C/D=1.5 and C/D=1.0. For C/D=0.5, a larger settlement of about 5 mm is observed.

6 INTERPRETATION

In most theoretical models face support is assumed to be independent of the overburden. This cannot be verified by the present investigation, where different pressures at three different overburdens are measured. With decreasing overburden, the necessary face support decreases. Similar results were reported in [4]. The measured pressures in the centrifuge tests agree well with the numeric analysis and some previously published data from the literature [5,6,7].

The observed soil deformation can be described using the failure mechanism as proposed by Horn [8], which consists of a prismatic wedge in front of the tunnel face, with a prismatic chimney on top, reaching the ground surface. Only for C/D=0.5, failure propagates to the surface, inducing great settlement. Failure surfaces for an overburden ratio of C/D=1.0 and C/D=1.5 propagate until 50mm and 100mm below the surface, respectively.

7 CONCLUSIONS

The support pressure is dependent on the overburden. While centrifuge tests and numeric analysis are in good agreement, some theoretical models overestimate the support pressure. At an overburden ratio of C/D=0.5, failure surfaces propagate until ground surface and induce large settlement.

REFERENCES

[1] Mair, R.J., Taylor, R.N.: Bored tunnelling in the urban environment. In: Proceedings of the 14th International Conference on Soil Mechanical and Foundation Engineering, Hamburg, vol. 4, pp. 2353–2385 (1997)

[2] White, D.J., Take, W.A., Bolton, M.D.: Soil deformation measurement using particle image velocimetry (PIV) and photogrammetry. Géotechnique 53(7), 619–631 (2003)

[3] White, D.J., Take, W.A., Bolton, M.D.: Measuring soil deformation in geotechnical models digital images and PIV analysis. In: 10th International Conference on Computer Methods and Advances in Geomechanics, Tuscan, Arizona, pp. 997–1002 (2001)

[4] Kolymbas, D.: Tunnelling and Tunnel Mechanics - A Rational Approach to Tunnelling. Springer, Berlin (2005)

[5] Leca, E., Dormieux, L.: Upper and lower bound solutions for the face stability of shallow circular tunnels in frictional material. Géotechnique 40(4), 581–606 (1990)

[6] Chambon, P., Corte, J.F.: Shallow tunnels in cohesionless soil: Stability of tunnel face. Journal of Geotechnical Engineering 120(7), 1148–1165 (1994)

[7] Vermeer, P., Ruse, N., Marcher, T.: Tunnel heading stability in drained ground. Felsbau 20(6), 8–24 (2002)

[8] Horn, M.: Horizontaler Erddruck auf senkrechte Abschlussflächen von Tunneln, Landeskonferenz der ungarischen Tiefbauindustrie, German translation by STUVA, Düsseldorf (1961) (in German)

APPLICATIONS TO GEOTECHNICAL PROBLEMS OF A MICROMECHANICAL MODELING OF DAMAGE

Séverine Levasseur[1,*], Frédéric Collin[1], Robert Charlier[1], and Djimédo Kondo[2]

[1]ArGEnCo – Geo[3]
Université de Liège, Chemin des chevreuils 1, 4000 Liège, Belgium
[*]e-mail: severine.levasseur@ulg.ac.be, web page: http://www.argenco.ulg.ac.be/geo3.php
[2]Institut d'Alembert
Université Pierre et Marie Curie (UPMC), 4 place Jussieu, 75252 Paris cedex 05, France
P.O. Box 808, L-231, Livermore, CA, 94551, USA
e-mail: djimedo.kondo@upmc.fr, web page: http://www.upmc.fr

Summary. The mechanical behaviors of geomaterials are significantly affected by the presence of voids or crack-like defects. The modeling of such behavior is classically performed by considering purely macroscopic or micromechanically-based damage models. In the perspective of applications in civil engineering or in geomechanics, we propose in this paper to evaluate a homogenization approach, based on Mori-Tanaka scheme, applied to micro-cracked materials. In order to provide an appropriate interpretation of the nonlinear behavior at macro-scale, the crack-induced damage is coupled to friction phenomena on closed cracks lips. The predictions of the coupled model are first analyzed on laboratory tests performed on Callovo-Oxfordian Clay. Then, they are extended to a numerical analysis of excavation damaged zones around tunnels.

Keywords: damage, excavation damaged zone, homogenization, laboratory tests, micro-macro behavior laws, tunnel drilling.

1 INTRODUCTION

A zone with significant irreversible deformations and changes in flow and transport properties (named Excavation Damaged Zone or EDZ) is expected to be formed around underground excavations in the deep geological layers considered for the high level radioactive waste disposal. Stress perturbations around the excavation could lead to a significant increase of the hydromechanical properties, related to diffuse and/or localized micro-cracks propagation in the material. The modeling of such behavior is classically performed by considering macroscopic or micromechanically-based damage models. Recent developments in the field of homogenization methods provide now physically and mathematically appropriate framework for the investigation of the behavior of micro-cracked media including the description of damage anisotropy-induced anisotropy, as well as cracks closure effects [1,2]. However, in the perspective of applications to civil engineering or geotechnical problems, like underground excavations, it is desirable to evaluate the different homogenization schemes by an analysis of their assumptions and of the macroscopic response that they predict. The purpose of the present study is to provide an analysis of a micromechanical damage model in order to give an appropriate interpretation of the nonlinear behavior at macro-scale under

R.I. Borja (Ed.): Multiscale and Multiphysics Processes in Geomechanics, SSGG, pp. 185–188.
springerlink.com

particular stress paths. At micro-scale, the considered model, based on Mori-Tanaka homogenization scheme, incorporates damage coupled with friction [1,3]. The predictions of the micromechanical model are first performed on classical laboratory tests in order to evaluate the capability of the model to reproduce behaviors of a clayey rock at macro-scale and compared to an existing elastoplastic model. Then, once the advantages of the micromechanical model are demonstrated, an application in the context of nuclear waste storage is proposed through the modeling of a tunnel drilling in Callovo-Oxfordian Clay with an assessment and a characterization of the excavation damaged zone.

2 MICROMECHANICAL MODEL

The micromechanical model is based on a thermodynamic potential Ψ^* consisting of purely elastic effects and inelastic effects due to damage d^r, crack opening β^r, and friction γ^r on each r-crack [1]:

$$\begin{aligned}\Psi^* &= \frac{1}{2}\boldsymbol{\Sigma} : \mathbb{S}^s : \boldsymbol{\Sigma} - \frac{1}{2}\sum_{r=1}^{p_c} \frac{w^r}{d^r}[H_0(\beta^r)^2 + H_1\boldsymbol{\gamma}^r \cdot \boldsymbol{\gamma}^r] \\ &+ \boldsymbol{\Sigma} : \sum_{r=1}^{p_c} w^r[\beta^r \boldsymbol{n}^r \otimes \boldsymbol{n}^r + (\boldsymbol{\gamma}^r \otimes \boldsymbol{n}^r)^s] \\ &+ \frac{1}{2}\boldsymbol{\Sigma} : \sum_{r=1}^{p_o} w^r\left(\frac{d^r}{H_0}\mathbb{E}^{2,r} + \frac{d^r}{2H_1}\mathbb{E}^{4,r}\right) : \boldsymbol{\Sigma}\end{aligned} \tag{1}$$

where $\boldsymbol{\Sigma}$ is the stress tensor, p_c is the number of closed cracks family, p_o the number of open cracks family, $\mathbb{E}^{2,r}$ and $\mathbb{E}^{4,r}$ are the tensors relative to the normal of crack $\boldsymbol{n}^r$, $\mathbb{S}^s$ is the inverse of the elastic constitutve matrices, i.e., $\mathbb{S}^s = (\mathbb{C}^s)^{-1}$, and $\mathbb{C}^s = 3k^s\mathbb{J} + 2\mu^s\mathbb{K}$. Also, H_0 and H_1 are material parameters defined as $H_0 = 3E^s/16(1 - \upsilon^s)$ and $H_1 = H_0(1 - \upsilon^s/2)$, and $(\boldsymbol{a} \otimes \boldsymbol{b})^s = \mathrm{symm}(\boldsymbol{a} \otimes \boldsymbol{b})$.

Then, the strain tensor is given by $\boldsymbol{E} = \partial\Psi^*/\partial\boldsymbol{\Sigma}$, and the first state law reads

$$\boldsymbol{E} = \mathbb{S}^s : \boldsymbol{\Sigma} + \sum_{r=1}^{p_c} w^r[\beta^r \boldsymbol{n}^r \otimes \boldsymbol{n}^r + (\boldsymbol{\gamma}^r \otimes \boldsymbol{n}^r)^s] + \sum_{r=1}^{p_o} w^r\left(\frac{d^r}{H_0}\mathbb{E}^{2,r} + \frac{d^r}{2H_1}\mathbb{E}^{4,r}\right) : \boldsymbol{\Sigma} \tag{2}$$

The thermodynamic force of damage is given by

$$F^{d^r} = \frac{\partial\Psi^*}{\partial d^r} = \frac{1}{2} \times \begin{cases} [H_0(\beta^r)^2 + H_1\boldsymbol{\gamma}^r \cdot \boldsymbol{\gamma}^r]/(d^r)^2 & \text{closed cracks} \\ \boldsymbol{\Sigma} : (\mathbb{E}^{2,r}/H_0 + \mathbb{E}^{4,r}/2H_1) : \boldsymbol{\Sigma} & \text{open cracks} \end{cases} \tag{3}$$

Considering the damage and friction criteria

$$f^r = F^{d^r} - (c_0 + c_1 d^r) \qquad \text{and} \qquad g^r = |\boldsymbol{F}^{\gamma^r}| + \mu_f F^{\beta^r} \tag{4}$$

with c_0 and c_1 being the damage resistance variables in J/m^2 and μ_f being the friction coefficient, the rates of d^r, β^r, and γ^r follow the consistency relations: $\dot{f}^r = 0$, $\dot{g}^r = 0$ [1,3]. The rate form of the coupled law is then given by $\dot{\boldsymbol{E}} = \mathbb{S}_t^{\mathrm{hom}} : \dot{\boldsymbol{\Sigma}}$, with

$$\mathbb{S}_t^{\mathrm{hom}} = \mathbb{S}^s + \sum_{r=1}^{p_c} w^r \frac{1}{H^r_{\gamma\beta d}} \frac{\partial g^r}{\partial\boldsymbol{\Sigma}} \otimes \frac{\partial g^r}{\partial\boldsymbol{\Sigma}} + \frac{1}{c_1}\sum_{r=1}^{p_o} w^r \frac{\partial f^r}{\partial\boldsymbol{\Sigma}} \otimes \frac{\partial f^r}{\partial\boldsymbol{\Sigma}} \tag{5}$$

where

$$\frac{\partial g^r}{\partial \boldsymbol{\Sigma}} = \boldsymbol{\nu}^r \otimes \boldsymbol{n}^r + \mu_f \boldsymbol{n}^r \otimes \boldsymbol{n}^r,$$
$$H^r_{\gamma\beta d} = \frac{1}{d^r}(H_1 + \mu_f^2 H_0) - \frac{1}{H^r_d}\Big(\frac{H_1\gamma^r \cdot \boldsymbol{\nu}^r + \mu_f H_0 \beta^r}{(d^r)^2}\Big)^2$$
$$H^r_d = \frac{1}{(d^r)^3}[H_0(\beta^r)^2 + H_1\gamma^r \cdot \gamma^r) + c_1\,,$$
$$\frac{\partial f^r}{\partial \boldsymbol{\Sigma}} = \boldsymbol{\Sigma} : \Big(\frac{1}{H_0}\mathbb{E}^{2,r} + \frac{1}{2H_1}\mathbb{E}^{4,r}\Big)\,,$$

with $\boldsymbol{\nu}^r$ being the friction rate direction.

3 APPLICATIONS

3.1 Analysis of Laboratory Test

To test the advantages and the drawbacks of the micromechanical model described in the previous section, an analysis of compression test performed on Callovo-Oxfordien clay (COX) (see [4]) is proposed.

Micromechanical model parameter identification by trial and error of this test leads to the results presented on Figure 1. Compared to the calibration of an elastoplastic model based on Van Eekelen criteria, we show that both these models permit to catch experimental measurements. On this loading example, micromechanical approach permits to provide similar results as those of macroscopic elastoplastic approaches with fewer parameters [3].

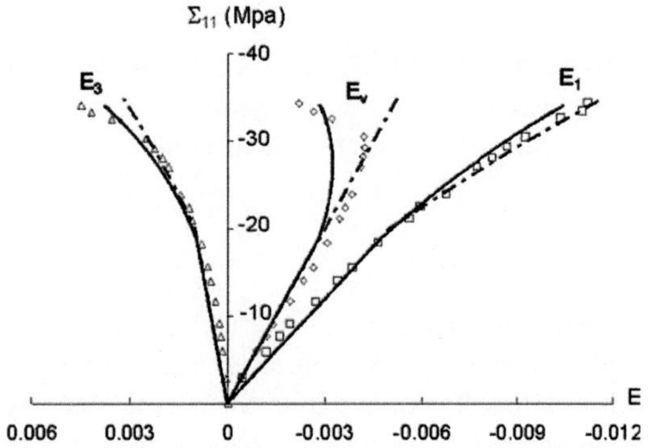

Fig. 1. Parameter calibration on compression test (symbols: experimental data; full lines: micromechanical model calibration; dash lines: elastoplastic model calibration)

3.2 Application to Tunnel Modeling

To go further in our tests on the proposed micromechanical model, a synthetic modeling of a tunnel excavation drilled in COX clay is proposed. Clay parameters identified in the previous section are used, and initial stresses are chosen isotropic and equal, for instance, to 30MPa. Tunnel excavation process consists in reducing the stress state from 30MPa to 0MPa. Applied on this in situ experiment, the micromechanical model provides an expected stress distribution and permits to define a realistic excavation damaged zone as shown on Figure 2, on which EDZ is defined from an equivalent damage variable D_{eq}, introduced by analogy with the equivalent stress Σ_{eq} [3]. Compared to elastoplastic model (for which EDZ is deduced from equivalent plastic stains $\varepsilon^{\mathrm{p}}_{\mathrm{eq}}$), the same EDZ size can be evaluated even if this EDZ does not characterize the same

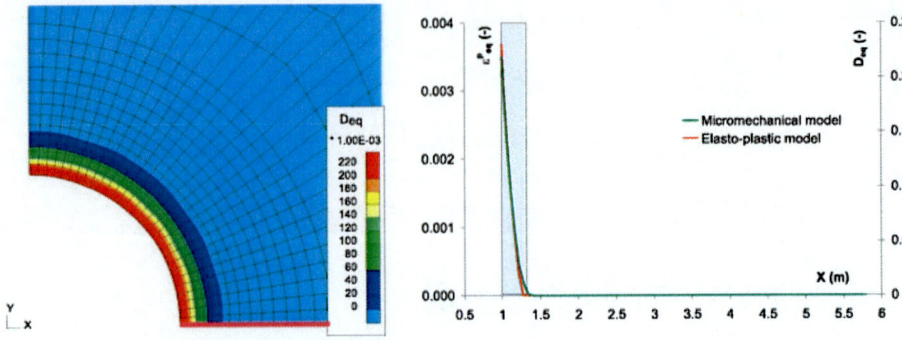

Fig. 2. Von Mises equivalent damage field according to micromechanical model; Comparison along the pink section between damage zone defined by micromechanical model (D_{eq}) and defined by elastoplastic model (ε^{P}_{eq} analogy between damage and plastic strains).

mechanical behavior. Based on the orientational distribution of the normal of damage (not shown in this article), an analysis of crack orientations in this zone shows that damage are linked to activated shear cracks on the tunnel borders. These shear cracks are oriented closed to $\pi/4 + \varphi/2$.

4 CONCLUSION

From the above, the micromechanical model seems to provide promising physical results, with appropriate interpretation of the EDZ, compared to elastoplastic models. This is a huge advantage from the perspective of evaluating transport properties in EDZ, like permeability evolution. To improve the micromechanical model, hydromechanical couplings as well as the dependency between initial stresses and damage–friction coupling will be considered in future analysis by taking advantage of the studies of Levasseur et al. [5,6].

REFERENCES

[1] Zhu, Q., Kondo, D., Shao, J., Pensée, V.: Micromechanical modelling of anisotropic damage in brittle rocks and application. Int. J. Rock Mech. Min. Sci. 45(4), 467–477 (2008)
[2] Dormieux, L., Kondo, D., Ulm, F.-J.: Microporomechanics. Wiley, Chichester (2006)
[3] Levasseur, S., Collin, F., Charlier, R., Zhu, Q., Kondo, D.: Damage modelling in geotechnics: micromechanical approach. In: Proc. of Eurock 2010, Lausanne, Switzerland, June 15-18, pp. 149–152 (2010)
[4] Abou-Chakra Guéry, A., Cormery, F., Shao, J., Kondo, D.: A micromechanical of elastoplastic and damage behavior of a cohesive geomaterial. Int. J. Solids Struct. 45(5), 1406–1429 (2008)
[5] Levasseur, S., Collin, F., Charlier, R., Kondo, D.: On a class of micromechanical damage models with initial stresses for geomaterials. Mechanics Research Communications 37, 38–41 (2010)
[6] Levasseur, S., Collin, F., Charlier, R., Kondo, D.: A two scale anisotropic damage model accounting for initial stresses in microcracked materials. Engineering Fracture Mechanics (2011), doi:10.1016/j.engfracmech.2011.03.009

NONLINEAR MICROMECHANICAL MODELING OF HYGROTHERMAL EFFECTS ON STRUCTURAL BIOBASED COMPOSITE MATERIALS

Wil V. Srubar III[1] and Sarah L. Billington[2]

[1] Department of Civil and Environmental Engineering
Stanford University
MC: 4020, Building 540, Stanford, CA 94305-4020
e-mail: wsrubar@stanford.edu
[2] 473 Via Ortega, Rm 285A, Stanford, CA 94305-4020
e-mail: billington@stanford.edu, web page: http://www.stanford.edu/~billingt/

Summary. Biobased composite materials fabricated from bacterial poly(β-hydroxybutyrate)-*co*-poly(β-hydroxyvalerate) (PHBV) resins reinforced with natural oak wood flour are being investigated as viable replacements to traditional, ecologically insensitive construction materials. The material properties achievable by these composites, which biodegrade anaerobically under specific conditions, are comparable to that of dimensional lumber; however, natural fiber composites remain particularly susceptible to hygrothermal degradation of the fiber–matrix interface because of the incongruent hydrophobicity of the biopolymer and the hydrophilicity of the cellulosic fibers.

A temperature-driven Fickian sorption–diffusion model is being employed to characterize the transient moisture absorption behavior of the wood flour composites. Constitutive models are being developed to incorporate volumetric swelling and fiber–matrix interfacial degradation due to moisture uptake. Simple, nonlinear micromechanical models using assumptions of isotropy can be employed to predict short-fiber composite stiffness and strength under combined environmental and mechanical loads. Preliminary experimental tests are being conducted to calibrate such models and to validate finite element simulations for combined mechano-sorptive composite behavior.

Keywords: biobased composites, diffusivity, predictive models, water absorption.

1 INTRODUCTION

Increased environmental awareness during the last decade has prompted the development of natural fiber composite material replacements for synthetic composites and plastics [1]. Lignocellulosic fibers are considered exceptional reinforcing materials for polymers due to such favorable attributes as rapid renewability, low specific density, high specific strength and stiffness, biodegradability, worldwide abundance, and potentiality for global economic viability [2, 3]. The first generation of natural fiber plastic composites (NFPCs), now in widespread application, use natural fibers as reinforcement in petroleum-based resins

R.I. Borja (Ed.): Multiscale and Multiphysics Processes in Geomechanics, SSGG, pp. 189–192.
springerlink.com

(e.g., HDPE, PP, etc.); however, the manufacture of these partial or hybrid biobased composites remains remarkably energy intensive, and landfills continue to be the sole channel for disposal. Thus, there exists a resounding need for the development of fully biodegradable composite materials derived from renewable, non-petroleum-based resources.

To address this need, randomly oriented short oak wood flour fibers were used as reinforcement in a poly(β-hydroxybutyrate)-*co*-poly(β-hydroxyvalerate) (PHBV) resin in the fabrication of true biobased composite materials. PHBV is a completely biodegradable and highly hydrophobic thermoplastic aliphatic microbial polyester produced via bacterial fermentation; it exhibits remarkably similar properties to synthetic plastics [4].

2 MICROMECHANICAL MODELING

The macromechanical properties of biobased composites are dictated by a) the properties of their individual constituents; and b) the micromechanics of the fiber–matrix interfacial bond. Previous research has proven that biobased composites can achieve similar mechanical properties to dimensional lumber; however, the composite is susceptible to hygrothermal degradation in elevated moisture and temperature environments due to the inherent hydrophilicity of natural fibers [5]. Moisture absorption leads to fiber swelling, plasticization of the matrix, volumetric/dimensional instability, and an overall loss of integrity of the micromechanical bond at the fiber-matrix interface.

A multiscale modeling approach is being developed to predict composite behavior under combined mechanical and environmental loads. Modeling the time-dependent mechano-hygrothermal behavior of biobased composites involves a two-stage process that couples a) moisture absorption and corresponding mechanical deformations and interfacial degradation; and b) total moisture content and its effect on the governing constitutive behavior. Simple composite theory models of moisture absorption, stiffness, and strength can be used to predict the fundamental behavior of the composites.

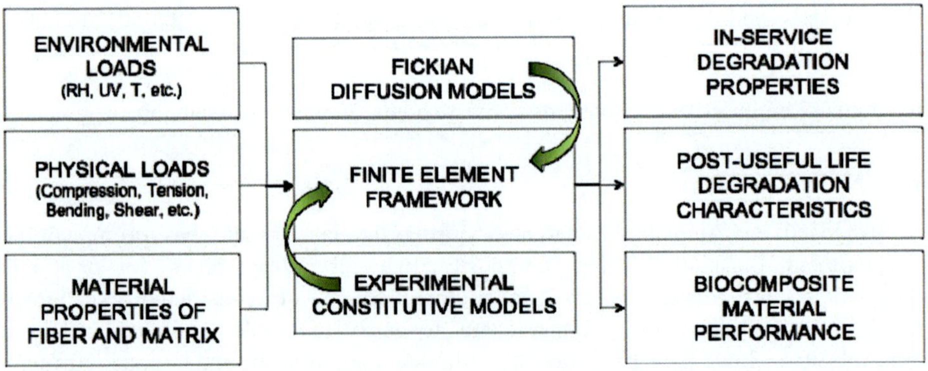

Fig. 1. Schematic of Proposed Multiscale Mechano-Hygrothermal Model Framework.

2.1 Moisture Absorption

The mechanistic behavior of water absorption in natural fiber reinforced biobased composites occurs in a highly complex way. Rigorous, proven models to describe these intricate mechanisms have not yet been developed [6]. Simplistically, moisture absorption processes in polymer composites can be described by Fick's second law of diffusion; however, rates of moisture uptake has been shown to depend upon a number of parameters: temperature, fiber volume fraction, fiber particle size, orientation, and vapor pressure differentials [7]. Thus, the diffusion coefficient, D, can be time-dependently characterized:

$$\frac{M_t}{M_\infty} = 1 - \frac{8}{\pi^2}\sum_{n=0}^{\infty}\frac{1}{(2n-1)^2} exp\left\{\frac{-D(T(t), RH(t), v, \lambda, \theta, \upsilon)(2n-1)^2\pi^2 t}{h^2}\right\} \tag{1}$$

where M_t and M_∞ are the moisture content at time t and at equilibrium, respectively. T is the ambient temperature, RH is the relative humidity, v, λ, and θ are the volume fraction, aspect ratio, and orientation of the wood fibers, respectively, υ is the water vapor pressure differential, and h is the minimum thickness dimension.

2.2 Composite Stiffness and Strength

Experimental data obtained from fundamental tensile tests can be compared with the most commonly used theoretical models for stiffness and strength. The Halpin-Tsai and Tsai-Pagano equations were developed to predict the longitudinal (E_L) and transverse (E_T) tensile modulus of aligned short wood fiber reinforced biopolymeric composites [8].

$$E_L = E_m \frac{1 + 2\lambda\eta_L v_f}{1 - \eta_L v_f} \tag{2}$$

$$E_T = E_m \frac{1 + 2\eta_T v_f}{1 - \eta_T v_f} \tag{3}$$

where E_m is the matrix tensile modulus, and η_L and η_T are given by the following:

$$\eta_L = \frac{(E_f / E_m) - 1}{(E_f / E_m) + 2\lambda} \tag{4}$$

$$\eta_T = \frac{(E_f / E_m) - 1}{(E_f / E_m) + 2} \tag{5}$$

The Nicolais and Nicodemo composite strength prediction model is used to evaluate the tensile strength of randomly oriented short fiber composites. This model assumes no adhesion between the matrix and filler; thus, it can be parametrically calibrated to encapsulate a composite's strength dependence upon the fiber volume fraction and degree of hygrothermal degradation, namely, on the moisture-induced plasticization at the fiber–matrix interface [8].

2.3 Mechano-Hygrothermal Constitutive Behavior

Combined mechanical and hygrothermal stress states influence rates of diffusivity and dimensional stability. Coefficients of thermal and hydro expansion can be derived from fundamental mechanical characterization and sorption–diffusion experiments. At any given combined state of loading, the internal strains can be calculated.

Using assumptions of material isotropy and plane stress, simple linear composite theory constitutive models can be employed to predict the linear-elastic behavior of the composite. To predict the long-term behavioral response, nonlinear constitutive models based on time-dependent exposure, matrix cracking, and mechanical load fluctuations must be developed along with finite element simulations to fully understand and adequately model the composite's transient mechano-sorptive behavior.

3 CONCLUSIONS

Biobased composites exhibit desirable mechanical properties but are prone to hygrothermal degradation due to the inherent hydrophilicity of natural fibers. Transient moisture uptake can be modeled using a modified Fickian diffusion law. Composite properties can be predicted using Halpin-Tsai and Nicolais and Nicodemo models for stiffness and strength, respectively. Further systematic experimentation and mechanical testing are essential for further calibration of micromechanical models and finite element simulations.

REFERENCES

[1] Gamstedt, E., Almgren, K.: Natural fibre composites – with special emphasis on effects of the interface between cellulosic fibres and polymers. In: Proceedings of the 28th International Symposium on Materials Science (2007)

[2] Athijayamani, A., et al.: Effect of moisture absorption on the mechanical properties of randomly oriented natural fibers/polyester hybrid composites. Materials Science and Engineering A 517, 344–353 (2009)

[3] Jacob, M., Varughese, K., Thomas, S.: Water sorption studies of hybrid biofiber-reinforced natural rubber composites. Biomacromolecules 6, 2969–2979 (2005)

[4] Dufresne, A., Dupeyre, D., Paillet, M.: Lignocellular flour-reinforced poly(hydroxybuturate-co-valerate) Composites. J. of Applied Polymer Science 87, 1302–1315 (2003)

[5] Christian, S.: Mechanical Characterization and Structural Assessment of Biocomposites for Construction. Ph.D. Dissertation, Stanford University (2009)

[6] Steckel, V., Clemons, C., Thoemen, H.: Effects of material parameters on the diffusion and sorption properties of wood-flour/polypropylene composites. J. of Applied Polymer Science 103, 752–763 (2007)

[7] Wan, Y., et al.: Mechanical, moisture absorption, and biodegradation behaviours of bacterial cellulose fibre-reinforced starch biocomposites. Composites Science and Technology 69, 1212–1217 (2009)

[8] Singh, S., Mohanty, A.: Wood fiber reinforced bacterial bioplastic composites. Composites Science and Technology 67, 1753–1763 (2007)

SIMULATION OF JSC-1A LUNAR SOIL SIMULANTS WITH NEW ALGORITHMS DEVELOPED FOR REALISTIC SOIL PARTICLE GEOMETRIES

S.J. Lee[1], Y.M.A. Hashash[1], R.A. Wilkinson[2], and J.H. Agui[2]

[1] Department of Civil and Environmental Engineering
University of Illinois at Urbana-Champaign
205 North Mathews Avenue, Urbana, IL 61801
e-mail: lee225@illinois.edu, hashash@illinois.edu, web page: http://www.illinois.edu
[2] NASA Glenn Research Center
21000 Brookpark Road, Cleveland, OH 44135
e-mail: Allen.Wilkinson@grc.nasa.gov, Juan.Agui@grc.nasa.gov,
web page: http://www.grc.nasa.gov

Summary. This article reviews novel algorithms for the Discrete Element Method (DEM) with polyhedral particles to greatly improve the efficiency in particle contact detection which is the most time-consuming part of a simulation. The implemented algorithms are used in simulations designed to reproduce experimental tests on angular JSC-1A lunar regolith stimulant particles.

Keywords: discrete element method, JSC-1A lunar soil stimulant, polyhedral particles.

1 INTRODUCTION

Discrete element simulation of lunar regolith and simulant is of significant interest to establish a long-term presence on the Moon. With the recent increase of the computing power in personal computers, the realistic discrete element modeling of the highly angular lunar soil grains becomes more feasible. In this article, several novel algorithms, implemented in DEM code BLOKS3D, to greatly enhance the performance of large scale polyhedral discrete element analysis, are reviewed. This article then describes the DEM simulation to reproduce the experimental tests on JSC-1A lunar soil simulant [1].

2 DEM SIMULATION WITH POLYHEDRAL PARTICLES

The use of polyhedral particles in DEM simulations allows for a more accurate representation of angular particles such as crushed rock, sands, and other types of angular particles. However, polyhedral particles contacts detection are computationally significantly more expensive than that of spherical or elliptical particles. A DEM code, BLOKS3D, has been developed for this purpose [2], which is used in this article. The code has several novel algorithms to enhance the performance of large scale simulations including:

Neighbor search algorithm: The two-level-search, an enhanced version of neighbor search algorithm of 3DEC [3], has been implemented in the code. The first level of search decomposes the space of interest into Eulerian cubic boxes, and makes

R.I. Borja (Ed.): Multiscale and Multiphysics Processes in Geomechanics, SSGG, pp. 193–196.
springerlink.com

the preliminary list of the neighboring particles that occupy the same box. The second level uses the bounding sphere of each particle to see if it is in contact, thus a more refined contact list is made.

Contact detection method: The shortest link method (SLM) [4], a common plane (CP) based contact detection method, has been implemented for a fast and detailed contact detection. In the conventional CP method, a large number of iterations are necessary to find a CP which defines the maximum gap between the particle pairs. In SLM, the link having the shortest distance between particles is sought because the perpendicular bisector plane to the link is the CP as illustrated in Figure 1. Finding the shortest link requires only a few iterations and is up to 17 times faster than the conventional CP identification method [4].

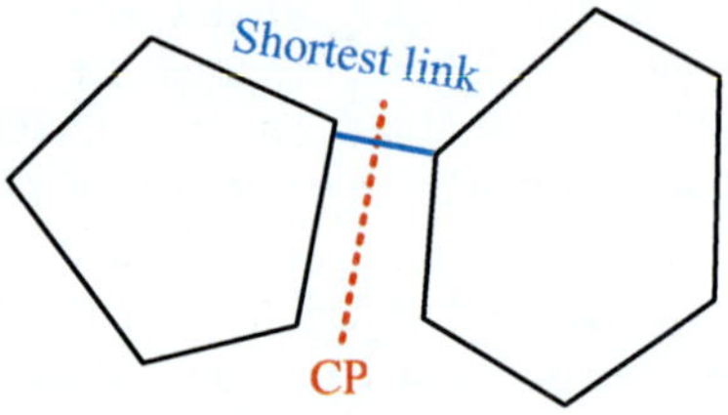

Fig. 1. The shortest link between two particles and the corresponding CP [4]

3 DEM SIMULATION OF EXPERIMENTAL TESTS ON JSC-1A

3.1 Experimental Tests on JSC-1A Lunar Regolith Simulant

Lunar regolith has distinct properties due to the weathering process on the Moon. Lunar regolith simulants are used as a substitute in experiments because of scarcity of actual regolith and are characterized by highly angular particles [5]. Figure 2 shows the experimental test setup devised by Agui and Bucek [1] to study the mechanical response of JSC-1A. A scoop is pushed into a 305 mm (width) × 330 mm (length) × 140 mm (height) rectangular soil bin. For a deep soil bed test, a deeper bin with a height of 292 mm has been adopted. The force sensor attached to the scoop measures the resistance forces in the reference coordinates y and z. Table 1 shows the parameters and the range of values considered in the tests.

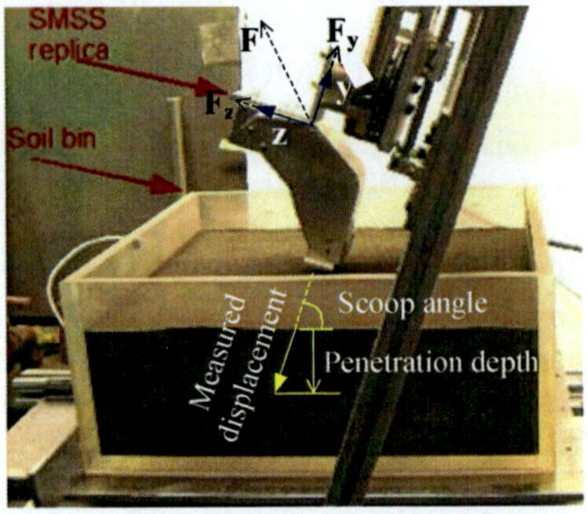

Fig. 2. Experimental setup [1]

Table 1. Parameters considered in tests on JSC-1A [1]

Parameters	Range of values
Soil bed depth (cm)	6.152 ~ 6.304, 9.053 ~ 11.285, 24.006 ~ 24.455
Scoop Velocity (mm/s)	1.270, 3.175, 6.350, 12.700
Scoop angle (°)	50, 60, 70
Bulk density (kg/m^3)	1576 ~ 1655, 1814 ~ 1871, 2022 ~ 2039
Penetration Depth (cm)	0.376 ~ 1.098, 3.058 ~ 4.308

3.2 DEM Particle Library Development

A new set of particle shapes were developed for use in the DEM particle library based on the unique shapes of JSC-1A. Scanning electron microscope photos of JSC-1A are used to establish a set of characteristic shapes as illustrated in Figure 3. The shapes were then simplified and approximated to corresponding polyhedral particle shapes as also shown in Figure 3.

Fig. 3. Scanning electron microscope images of JSC-1A and DEM particle library development

3.3 Grain Size Distribution and Other DEM Parameters

The size of the particles used in a simulation determines both the number of particles and the time step. The grain size distribution of JSC-1A is shown in Figure 4 with a minimum particle size of around 0.005 mm [6]. Significant computational cost can be incurred if particle sizes are adopted as they are in the DEM model. Figure 4 also shows a grain size distribution curve whereby the particle sizes are scaled up 100 times. However, the maximum grain size from this grain size distribution is 200 mm which is too large for the modeled bin. Therefore, a revised grain size distribution curve has been employed so that a total of about 17000 particles with a time step of 1.36×10^{-5} sec are used in the DEM simulation. Some of the DEM parameters and geotechnical properties, such as contact stiffness, friction angle have been adopted from [1, 6, 7]. Global damping was neglected, and contact damping was calibrated from the selected experiments.

3.4 DEM Simulation Result

Figure 5 shows the comparisons of DEM simulation results with a corresponding test data for soil bed depth: 6.152cm, bulk density: 1642kg/m^3, scoop velocity: 6.35mm/sec, scoop angle: 70°, and penetration depth: 3.058cm. The vector sum of the applied forces in the reference coordinates y and z is used for comparison. Results from two different trials using two different random particle initializations in the soil bin are shown. The computed forces show trends similar to those from experiments. In contrast to the continuous plots from the experiments, limited stick–slip force fluctuations are computed in the simulation. This is caused by the larger size of particles adopted in the DEM simulations. Nevertheless, there is an overall good agreement with the experimental values. Simulations of other configurations show similar trends.

4 CONCLUDING REMARKS

It is shown that the DEM approach with polyhedral particles is reliable to reproduce the experimental tests on JSC-1A lunar soil simulant. Although larger size of particles have been adopted to keep the computational resources manageable, the simulations show comparable results to those measured in the experiments on the whole.

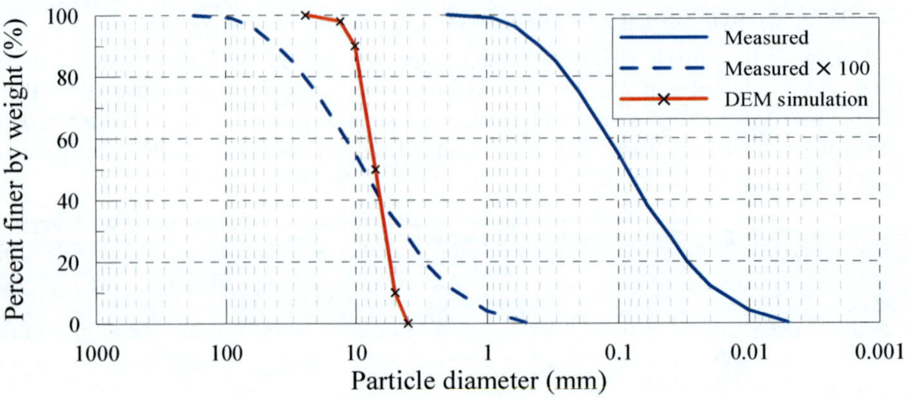

Fig. 4. The grain size distribution

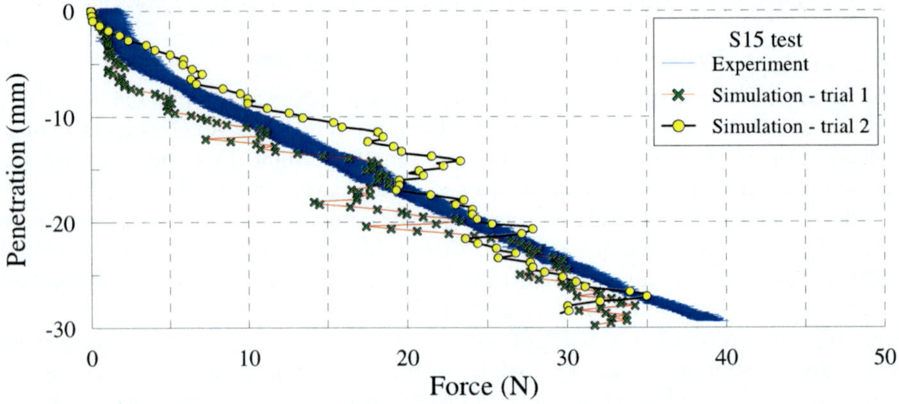

Fig. 5. Simulation results

REFERENCES

[1] Agui, J., Bucek, M.: Surveyor replica excavation task bearing test report. Technical report, NASA Glenn Research Center (GRC), Cleveland, OH (2007)

[2] Zhao, D., Nezami, E., Hashash, Y., Ghaboussi, J.: Three-dimensional discrete element simulation for granular materials. Eng. Computation 23(7), 749–770 (2006)

[3] Itasca Consulting Group, 3-dimensional distinct element code (3DEC) User's Manual (1998)

[4] Nezami, E., Hashash, Y., Zhao, D., Ghaboussi, J.: Shortest link method for contact detection in discrete element method. Int. J. Numer. Anal. Methods Geomech. 30(8), 783–801 (2006)

[5] Battler, M., Richard, J., Boucher, D., Spray, J.: Developing an anorthositic lunar regolith simulant. In: 37th Annual Lunar and Planetary Science Conference, League City, TX, March 13-17 (2006)

[6] Zeng, X., He, C., Oravec, H., Wilkinson, A., Agui, J., Asnani, V.: Geotechnical properties of JSC-1A lunar soil simulant. J. Aerosp. Engrg. 23(2), 111–116 (2010)

[7] Nezami, E., Hashash, Y., Zhao, D., Ghaboussi, J.: Simulation of front end loader bucket–soil interaction using discrete element method. Int. J. Numer. Anal. Meth. Geomech. 31, 1147–1162 (2007)

A Workshop Program

International Workshop on Multiscale and Multiphysics Processes in Geomechanics
Stanford University Campus, Building 260, Room 113
June 23–25, 2010

Organizing Committee

R. I. Borja	Civil & Environmental Engineering, Stanford (Chair)
E. M. Dunham	Geophysics, Stanford
E. Kuhl	Mechanical Engineering, Stanford
J. A. White	Lawrence Livermore National Laboratory

International Scientific Committee

J. S. Chen, Los Angeles, U.S.A.
W. Ehlers, Stuttgart, Germany
A. Gens, Barcelona, Spain
L. Laloui, Lausanne, Switzerland
X. B. Lu, Beijing, China
H. Mühlhaus, St. Lucia, Australia
M. Murad, Petropolis-RJ, Brazil
A. P. S. Selvadurai, Montreal, Canada
C. Tamagnini, Perugia, Italy
G. Viggiani, Grenoble, France
W. Wu, Vienna, Austria
H. S. Yu, Nottingham, U.K.

Sponsors

U.S. National Science Foundation, Geomechanics and Geomaterials Program
John A. Blume Earthquake Engineering Center, Stanford University

Time	Wednesday, June 23	Thursday, June 24	Friday, June 25
8:00	**Registration**		
8:15	**Opening Remarks**	**Registration**	
8:30	G. Viggiani	A.P.S. Selvadurai	X.B. Lu
8:52	D. Peric	M.A. Murad	J.P. Morris
9:14	M. Gutierrez	L. Laloui	H.S. Bhat
9:36	T. Triantafyllidis	P. Arduino	H. Noda
9:58	A. Rechenmacher	C.D. Foster	E.M. Dunham
10:20	**Break**	**Break**	**Break**
10:40	F.-J. Ulm	A. Gens	J.R. Rice
11:02	C. Dascalu	N. Lu	N. De Paola
11:24	J. Eberhardsteiner	C. Tamagnini	L. do N. Guimarães
11:46	A.M. da Costa	G. Hofstetter	I. Einav
12:08	J.E. Andrade	W. Ehlers	C.G. Sammis
12:30	**Lunch**	**Lunch**	**Lunch**
14:00	R. Chambon	R. de Borst	Y. Fialko
14:22	H. Mühlhaus	J.S. Chen	O. Vorobiev
14:44	W. Wu	G. Wells	C. Santamarina
15:06	R.A. Regueiro	R.I. Borja	**Summary**
15:28	**Break**	**Break**	
15:48	H.-S. Yu	J.A. White	
16:10	A. Niemunis	A. Tordesillas	
16:32	**Posters, Building 540**	**SLAC Tour**	
19:30	**Cocktails at Pampas Palo Alto (Brazilian)**		
20:00		**Banquet at MacArthur Park Restaurant**	

Wednesday, June 23, 2010

Multiscale Processes 1

Chair: R.A. Regueiro

8:30 am	Mechanisms of localized deformation in geomaterials studied through full-field measurement techniques **G. Viggiani**, P. Bésuelle, S.A. Hall. J. Desrues
8:52 am	Meso-scale evolution of strain localization observed in plane strain experiments on clay **D. Peric**, M. Gutierrez
9:14 am	Micromechanical study of strain localization and failure in the direct shear test of granular soils **M. Gutierrez**, J. Wang
9:36 am	Shear banding in geomaterials in plane strain: Physical and analytical model **T. Triantafyllidis**, L. Röchter, A. Niemunis
9:58 am	Experimental evidence of structural development inside shear bands in sands **A.L. Rechenmacher**, S. Abedi, I. Faoro

10:20 am	**Break**

Multiscale Processes 2

Chair: A.L. Rechenmacher

10:40 am	Scratch test model for micro- and macro-fracture properties of geomaterials A.-T. Akono, **F.-J. Ulm**
11:02 am	A two-scale damage law for creeping rocks **C. Dascalu**, B. François
11:24 am	Strength of geomaterials: Multiscale theories and experiments at appropriate problem-dependent length scales B. Pichler, C. Hellmich, **J. Eberhardsteiner**
11:46 am	Geomechanics applied to the well design through salt layers in Brazil: A history of success **A.M. Costa**, E. Poiate Jr., C.S. Amaral, A. Pereira, L.F. Martha, M. Gattass, D. Roehl
12:08 pm	Multiscale modeling of granular matter: A hierarchical scheme **J.E. Andrade**, C.F. Avila

12:30 pm	**Lunch**

Wednesday, June 23, 2010

Multiscale Processes 3

Chair: G. Viggiani

2:00 pm	Bifurcations from the microcale to the macroscale: A new paradigm for modeling and experiments **R. Chambon**
2:22 pm	Large-scale shear banding in extension **H. Mühlhaus**, L. Olsen-Kettle, J. Shi, L. Moresi
2:44 pm	On scale effect in growth of plant roots **W. Wu**, S. Ladjal
3:06 pm	Coupling discrete elements and micropolar continuum through an overlapping region in one dimension **R.A. Regueiro**, B. Yan

3:28 pm	**Break**

Multiscale/Multiphysics Processes

Chair: J.E. Andrade

3:48 pm	Sand behaviour under principal stress rotation: Observations and modelling **H.-S. Yu**, X. Li, Y. Cai
4:10 pm	Paraelastic deformations in hypoplasticity **A. Niemunis**, F. Prada

4:32 pm	**Poster Sessions, Blume Center Patio**

7:30 pm	**Cocktails at Pampas Palo Alto (Brazilian)**

Thursday, June 24, 2010

Multiphysics Processes 1

Chair: N. Lu

8:30 am	The role of modeling and simulations in estimating multiscale effective permeability **P.A. Selvadurai**, A.P.S. Selvadurai
8:52 am	The scale-up problem for ion transport in kaolinite clays including pH-dependence I. Igreja, **M.A. Murad**, S. Lima, C. Boutin
9:14 am	Biogrout propagation in soils **L. Laloui**, S. Fauriel
9:36 am	Modeling multi-scale flow using the material point method **P. Arduino**, P. Mackenzie-Helnwein, G.R. Miller
9:58 am	Numerical modeling of poromechanics in continuous and localized regions **C.D. Foster**, T.M. Nejad

10:20 am	Break

Multiphysics Processes 2

Chair: J.A. White

10:40 am	Coupled analysis of double porosity swelling clays **A. Gens**, L. do N. Guimarães, M. Sánchez, B. Valleján
11:02 am	Interpreting the "collapsing" behavior of unsaturated soil by effective stress principle **N. Lu**
11:24 am	Multiphysics modeling of electrokinetic phenomena in unsaturated fine-grained soils **C. Tamagnini**, C. Jommi, F. Cattaneo, D. Salciarini
11:46 am	A numerical model for geotechnical problems involving partially saturated soils **G. Hofstetter**, M. Pertl, M. Hofmann
12:08 pm	Experimental and computational issues in the mechanics of multi-physical unsaturated soil **W. Ehlers**, O. Acvi

12:30 pm	Lunch

Thursday, June 24, 2010

Computational Modeling 1

Chair: E. Kuhl

2:00 pm	Isogeometric failure analysis **R. de Borst**, T.J.R. Hughes, M.A. Scott, C.V. Verhoosel
2:22 pm	Micro-crack informed multi-scale damage model: Theory and computation X. Ren, **J.S. Chen**, J. Li
2:44 pm	Adaptive methods for multiphase flow through porous media **G. Wells**, H. Narayanan
3:06 pm	Circumventing the volume and contact constraints in multiphysics finite element simulations **R.I. Borja**, J.A. White, F. Liu

3:28 pm	**Break**

Computational Modeling 2

Chair: C.D. Foster

3:48 pm	Scalable preconditioning techniques for fully-coupled hydromechanical models **J.A. White**, R.I. Borja
4:10 pm	When failure is an option: Rewiring granular networks **A. Tordesillas**

4:32 pm	**Tour of SLAC National Accelerator Laboratory and Stanford Synchrotron Radiation Lightsource**

8:00 pm	**Banquet at MacArthur Park Restaurant, Palo Alto**

Friday, June 25, 2010

Fluids, Fault Zones and Earthquakes 1

Chair: N. De Paola

8:30 am	Formation of layered fracture and outburst by percolation **X.B. Lu**, X.H. Zhang, S.Y. Wang
8:52 am	Permeability evolution in fractured rock in response to mechanical and chemical processes **J.P. Morris**
9:14 am	Experimental and numerical modeling of the effect of off-fault damage on dynamic earthquake ruptures **H.S. Bhat**, A.J. Rosakis, C.G. Sammis
9:36 am	Earthquake sequence calculations with dynamic weakening mechanisms: Statically strong but dynamically weak fault and low interseismic shear stress **H. Noda**, N. Lapusta, J.R. Rice
9:58 am	Earthquake ruptures on rough faults **E.M. Dunham**, J.E. Kozdon, D. Belanger, L. Cong

10:20 am	**Break**

Fluids, Fault Zones and Earthquakes 2

Chair: E.M. Dunham

10:40 am	Stability and localization of rapid shear in fluid-saturated fault gouge **J.R. Rice**, J.W. Rudnicki
11:02 am	Fault lubrication and earthquake propagation in thermally unstable rocks **N. De Paola**, T. Hirose, T. Mitchell, G. Di Toro, C. Viti, T. Shimamoto
11:24 am	Coupled hydro-mechanical and chemical analysis of fault reactivation during CO2 injection in carbonate reservoirs **L. do N. Guimarães**, L.C. Pereira, I.F. Gomes, A. Gens
11:46 am	Heat convection in dry fault gouges P. Rognon, **I. Einav**
12:08 pm	Multi-scale fracture networks in geomechanics **C.G. Sammis**

12:30 pm	**Lunch**

Friday, June 25, 2010

Fluids, Fault Zones and Earthquakes 3

Chair: I. Einav

2:00 pm	Coseismic and early post-seismic deformation due to the April 4, 2010 M7.2 Durango (Mexico) earthquake from space geodetic and field observations **Y. Fialko**
2:22 pm	Equivalent continuum modeling for wave propagation in discontinuous media **O. Vorobiev**
2:44 pm	Emergent phenomena in energy geotechnology **C. Santamarina**

3:06 pm	**Summary**

Poster Sessions, Wednesday, 23 June 2010, 4:40-6:00 pm, Blume Center Patio

P1	Vortex structures inside shear bands in sands **S. Abedi**, A. Rechenmacher
P2	Simulation of hydration and elastic properties of montmorillonite using molecular dynamics **D. Ebrahimi**, R.J.M. Pellenq, A.J. Whittle
P3	Emergent chemo-hydro-mechanical phenomena in carbon geological storage **D.N. Espinoza**, J.C. Santamarina
P4	Implementation of multiscale models in a probabilistic framework for performance-based durability engineering **M.M. Flint**, S.L. Billington
P5	Centrifuge model test and numerical analysis on face stability of shallow tunnels **G.D. Idinger**, P. Aklik, X. Song, W. Wu
P6	Simulation of JSC-1A lunar soil simulants with new algorithms developed for realistic soil particle geometries **S.J. Lee**, Y.M.A. Hashash, R.A. Wilkinson, J.H. Agui
P7	Applications to geotechnical problems of the micromechanical modeling of damage **S. Levasseur**, F. Collin, R. Charlier, D. Kondo
P8	Dynamic fault rupture and propagation through a fixed finite element grid **F. Liu**, R.I. Borja
P9	Landslides and debris-flow induced static and dynamic loads on protective structures **C.M. Mast**, P. Mackenzie-Helnwein, P. Arduino, G.R. Miller
P10	Strength homogenization for cohesive-frictional composites: A linear comparison composite approach **J.A. Ortega**, B. Gathier, F.-J. Ulm
P11	Paraelasticity – a constitutive model for small strains A. Niemunis, **F. Prada**
P12	Nonlinear micromechanical modeling of hygrothermal effects on structural biobased composite materials **W.V. Srubar**, S.L. Billington
P13	Diffuse bifurcations of porous media under partially drained conditions **W.C. Sun**, J.E. Andrade
P14	Discontinuous pore fluid distribution under microgravity due to particle rearrangement **M. Xiao**, L.N. Reddi, S.L. Steinberg
P15	Multiphysical simulations of the freezing effects on soils Z. Liu, **X. Yu**

Tour of SLAC National Accelerator Laboratory and Stanford Synchrotron Radiation Lightsource, Stanford University, 24 June 2010

5:00 pm – 5:30 pm	Presentation by Dr. Piero Pianetta, Acting Director of SLAC's Synchrotron Radiation Laboratory
5:30 pm – 5:35 pm	Board Buses with Tour Guides

Tour with George Kuraitis

5:35 – 6:00 pm	Drive to SSRL main office. Meet with SSRL tour guide Mike Toney for tour
6:00 – 6:30 pm	Drive to Visitors Alcove, Klystron Gallery tour

Tour with Dan Miller

5:35 – 6:00 pm	Drive to Visitors Alcove, Klystron Gallery tour
6:00 – 6:30 pm	Drive to SSRL main office. Meet with SSRL tour guide Clyde Smith for tour
6:30 pm	Buses depart Main Gate
Websites	SLAC: http://www.slac.stanford.edu
	SSRL: http://ssrl.slac.stanford.edu

Reception at Pampas Restaurant, 23 June 2010

Cocktails

Classic Caipirinha, Gaucho, Alma, Gypsy, Samba-Rita, Sangria-Rio, Beijo Muitos, Cosmo do Diabo, Anjo de Paraiso

Ceviche, Lamb Meatball, Empanadas, Lamb Chops, Prawns, Baked Queso Fresco, Tenderloin, Mushroom Polenta

Banquet at MacArthur Park Restaurant, 24 June 2010

Mesquite-Grilled Sausages, Phylo Tartletts, MacArthur Park Mini Maryland Crab Cakes
Mac Arthur Park Salad
MacArthur Park's Famous Baby Back Pork Ribs Half Slab

Natural Barbeque Half Chicken
Fire Roasted Vegetable Kabob
Mud Pie

Ronaldo I. Borja, Piano
Two Etudes by Frédéric Chopin
A-flat major Op. 25 No. 1
E major Op. 10 No. 3

B Author Index

C List of Participants

Sara Abedi	University of Southern California
Jose Andrade	California Institute of Technology
Pedro Arduino	University of Washington
Harsha Bhat	University of Southern California
Ronaldo Borja	Stanford University
René Chambon	University of Grenoble
Jiun-Shyan Chen	University of California, Los Angeles
Qiushi Chen	Northwestern University
Cristian Dascalu	University J. Fourier, Grenoble
René de Borst	Eindhoven University of Technology
Nicola de Paola	University of Durham
Eric Dunham	Stanford University
Josef Eberhardsteiner	Vienna University of Technology
Davoud Ebrahimi	Massachusetts Institute of Technology
Wolfgang Ehlers	University of Stuttgart
Itai Einav	The University of Sydney
D. Nicolas Espinoza	Georgia Institute of Technology
Arvin Farid	Boise State University
Yuri Fialko	University of California, San Diego
Madeleine Flint	Stanford University
Craig Foster	University of Illinois at Chicago
William Fuentes	Karlsruhe Institute of Technology
Marcelo Gattass	Tecgraf/PUC-Rio
Antonio Gens	Technical University of Catalonia
Roozbeh Geraili	University of California, Berkeley
Leonardo do Guimarães	Federal University of Pernambuco
Marte Gutierrez	Colorado School of Mines
Guenter Hofstetter	University of Innsbruck

Zhiqiang Hu	Dalian University of Technology
Gregor Idinger	Universität für Bodenkultur Vienna
Ellen Kuhl	Stanford University
Lyesse Laloui	École Polytechnique Fédérale de Lausanne
Seung Jae Lee	University of Illinois
Severine Levasseur	Université de Liège
Xiaoyu Liu	Stanford University
Fushen Liu	Stanford University
Xiaobing Lu	Chinese Academy of Sciences
Ning Lu	Colorado School of Mines
Peter Mackenzie-Helnwein	University of Washington
Alvaro Maia	Petrobras
Carter Mast	University of Washington
Jörg Meier	Stanford University
Holger Meier	ExxonMobil
Steven Meier	ExxonMobil Research and Engineering Company
Daniela Menozzi	University of Rome La Sapienza
Euclides Mesquita	UNICAMP - University of Campinas Brazil
Joseph Morris	Schlumberger-Doll Research
Hans Mühlhaus	The University of Queensland
Marcio Murad	Laboratório Nacional de Computação Cientifica
Andrzej Niemunis	Karlsruhe University
Hiroyuki Noda	California Institute of Technology
Jose Ortega	Massachusetts Institute of Technology
Leonardo Pereira	Petrobras
Dunja Peric	Kansas State Univeristy
Luis Felipe Prada Sarmiento	Karlsruhe Institute of Technology
Helia Rahmani	Stanford University
Amy Rechenmacher	University of Southern California
Richard Regueiro	University of Colorado at Boulder
James Rice	Harvard University
Deane Roehl	Tecgraf/PUC-Rio
Charles Sammis	University of Southern California
Marcelo Sanchez	Texas A&M University
J. Carlos Santamarina	Georgia Institute of Technology

Pablo Sanz	Exxon Mobil
Paul Selvadurai	McGill University
Xiaoyu Song	Universität für Bodenkultur Vienna
Wil Srubar	Stanford University
WaiChing Sun	Northwestern University
Claudio Tamagnini	University of Perugia
Martin Tjioe	Stanford University
Antoinette Tordesillas	University of Melbourne
Theodoros Triantafyllidis	Karlsruhe Institute of Technology
Franz Ulm	Massachusetts Institute of Technology
Lillibelle Uriarte	MTA/LIRR New York
Gioacchino Viggiani	University of Grenoble
Oleg Vorobiev	Lawrence Livermore National Laboratory
Garth Wells	University of Cambridge
Joshua White	Lawrence Livermore National Laboratory
Wei Wu	Universität für Bodenkultur Vienna
Ming Xiao	California State University, Fresno
Hai-Sui Yu	University of Nottingham
Xiong Yu	Case Western Reserve University
Gregory Zarski	Stanford University
Degao Zou	Dalian University of Technology

Springer Series in Geomechanics and Geoengineering

Edited by W. Wu, R.I. Borja

Tejchman, J.
Shear Localization in
Granular Bodies with Micro-Polar
Hypoplasticity
316 p. 2008 [978-3-540-70554-3]

Iskander, M.
Modelling with Transparent Soils
331 p. 2010 [978-3-642-02500-6]

Tejchman, J. and Kozicki, J.
Experimental and Theoretical
Investigations of
Steel-Fibrous Concrete
289 p. 2010 [978-3-642-14602-2]

Iskander, M.
Behavior of Pipe Piles in Sand
267 p. 2010 [978-3-642-13107-3]

Wan, R., Alsaleh, M., and Labuz, J. (Eds.)
Bifurcations, Instabilities and Degradations
in Geomaterials
367 p. 2011 [978-3-642-18283-9]

Borja, R.I. (Ed.)
Multiscale and Multiphysics Processes in
Geomechanics
213 p. 2011 [978-3-642-19629-4]